INTRODUCTION TO
*the Solar Wind*

*A Series of Books in Astronomy and Astrophysics*
EDITORS: Geoffrey Burbidge
Margaret Burbidge

INTRODUCTION TO

# the Solar Wind

## John C. Brandt

NASA-Goddard Space Flight Center
Greenbelt, Maryland

*W. H. Freeman and Company*
San Francisco

*Indication of the author's affiliation with the
National Aeronautics and Space Administration
on the title page of this book does not constitute
endorsement of any of the material contained
herein by the NASA. The author is solely
responsible for any views expressed.*

Printed in the United States of America
Library of Congress Catalog Card Number: 75–89919
Standard Book Number: 7167 0328–9

1  2  3  4  5  6  7  8  9

*for Dorothy*

# Contents

# *Preface*

In the fall of 1959, I was privileged to hear E. N. Parker speak at the Yerkes Observatory on the topic "Solar Wind," a lecture in which he outlined his theory of the hydrodynamic expansion of the solar corona. At that time his theory was received with a great deal of skepticism, even disbelief. Today, only a decade later, the solar wind is an established astronomical fact and the object of many sophisticated and expensive space experiments. It has had an impact on many areas of astronomy and space science, such as the formation of ionic comet tails, stellar evolution and rotation, and the theory of general relativity. The solar wind has become an integral and natural part of solar physics and astronomy.

However, just as it was born into controversy, so today this atmosphere still persists to some extent, for the subject of solar wind physics is progressing quickly, and the picture is not always a clear one. Therefore, although this Introduction is intended to be broad-based, complete objectivity in all cases is not possible. Consider this treatise as somewhat akin to a description of the Grand Canyon near sunrise; the broad outline is clear, but many areas are still in shadows. The conscientious student will want to keep up with all the new information about the solar wind which is appearing in the latest journals even as I write these words.

This text is aimed at the advanced undergraduate and the beginning graduate student and is intended to provide a compact, comprehensive introduction with emphasis on basic principles and broad concepts. It is hoped that this monograph will also serve as an introduction for scientists in other fields as well as for technical and engineering workers involved in space and space-related efforts.

It is a pleasure to thank Drs. Stuart Jordan, Norman F. Ness, and John M. Wilcox for their interest, comments and assistance in the preparation of this manuscript. I am also indebted to numerous authors and institutions for their kindness and cooperation in supplying illustrations.

*Greenbelt, Maryland*
*May, 1969*
                                                        *John C. Brandt*

INTRODUCTION TO
*the Solar Wind*

# 1

# *Historical Introduction and Summary*

The solar wind is basically a proton-electron gas that streams past the earth with a mean velocity of 400-500 km/sec and a mean proton and electron density of about $5/cm^3$. Today the solar wind is an accepted fact, important in the discussion of many facets of astrophysics and space science. Based mainly on the existence of the zodiacal light the existence of an interplanetary medium was considered for centuries. While many of these early theories were erroneous, some were surprisingly close to the mark and therefore deserve brief mention.

## 1.1   Prehistory

A very early theory of the zodiacal light was outlined in 1672 by Cassini, who ascribed this faint radiation to an asymmetrical dust cloud around the sun (Figure 1.1). This hypothesis of scattering particles composed of dust has been open to a great deal of questioning. The hypothesis of scattering by electrons enjoyed considerable popularity in the early and middle 1950s— 300 years later—but was abandoned when it became apparent that the

2

FIGURE 1.1
The zodiacal light as seen from Mount Chacaltaya, Bolivia.
(Courtesy D. E. Blackwell and M. F. Ingham.)

electron densities required were far too high (see Section 1.2 for a brief discussion of the zodiacal light problem).

Early cometary studies were also concerned with an interplanetary resisting medium in connection with the forms of comet tails. Some of Bessel's pioneering work in 1835 was addressed to investigating Newton's and Enke's idea of a resisting "ether" to explain the shapes of comet tails as they were known at the time. Bessel concluded, however, that such a medium was not necessary, and it was over a hundred years later when conclusive evidence to the contrary was obtained (see the discussion of Hoffmeister's observations in Section 1.3). At present, an expanding resistive medium is required for the interpretation of the orientations of ionic comet tails (Section 4.1).

On September 1, 1859, the pioneering solar physicist Carrington carried out the first observation of an intense, white-light solar flare; independently Hodgson observed this same phenomenon. The flare was observed between 11:20 and 11:25 A.M., and a moderate geomagnetic disturbance was recorded at the same time. Approximately four hours after the following midnight, a great magnetic storm occurred, accompanied by intense aurorae that were observed in both hemispheres and at low latitudes. Carrington was fully aware of the possible implications of the sequence of events, but he was wary of suggesting a cause-and-effect relationship. Solar activity and the resultant production of aurorae, magnetic storms, etc., provided the motivation for many early studies of gas emission from the sun. Becquerel, for example, suggested sunspots as the solar source of auroral particles.

Sir Oliver Lodge, in an intriguing paper written in 1900 entitled "Sunspots, Magnetic Storms, Comet Tails, Atmospheric Electricity and Aurorae," wrote that magnetic storms were caused by "a torrent or flying cloud of charged atoms or ions." The velocity of the clouds responsible for these phenomena had been estimated by G. F. FitzGerald in 1892 (from time delays between central meridian passage of a sunspot and the associated magnetic storm) at "about 300 miles per second."

Lodge's and FitzGerald's ideas were not accepted at the time. The connection between geomagnetic activity and the solar cycle was considered coincidental by many people, including Lord Kelvin. However, these theories slowly gained recognition during the first half of the twentieth century, and by 1950 both Bartels (on the basis of geomagnetic evidence) and Biermann (from his studies of comet tails) had substantial evidence for a continuous solar corpuscular emission (see Section 1.3).

## 1.2 History

Many developments in a variety of fields contributed to the advent of the modern era in solar wind studies. In the early part of the twentieth century, these developments were almost entirely the result of studies of auroral and geomagnetic phenomena.

Birkeland's studies (*c.* 1903), including his auroral simulation with a laboratory terrella, led to many concepts which were accepted much later (see Figure 1.2 for a modern terella experiment). For example, he postulated solar-active regions that continuously emit material and, because of the solar rotation, produce magnetic storms with a twenty-seven-day period. He also concluded that the bombardment of the earth by solar material must be essentially continuous because there are almost always small magnetic storms. Compare Birkeland's conclusions with Bartels' views several decades later (see below).

Stoermer's work on aurorae was inspired by Birkeland's studies. Stoermer's computations of the paths of charged particles (originally thought to be electrons) in a dipole field were most applicable to a beam containing particles of one sign; such beams were seriously considered by a number of workers,

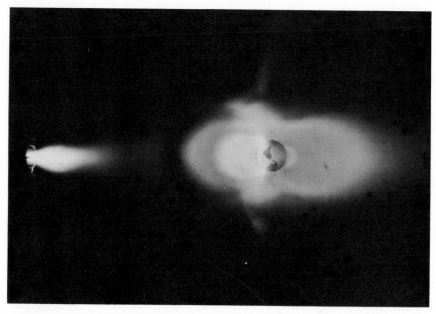

FIGURE 1.2

A modern version of the terella experiment illustrating the interaction of the solar wind with the earth's magnetic field. The plasma source is a magnetoplasmadynamic thruster (right). The trapped radiation belts are to scale. (Courtesy G. R. Seikel, NASA Lewis Research Center, Cleveland, Ohio.)

including Chapman, as the cause of aurorae. However, in 1911 Schuster voiced a fundamental objection: electrostatic repulsion would rapidly disperse a beam with particles of one sign. Hence, Lindemann, in 1919, first suggested a cloud of ionized matter with electrical neutrality as the cause of aurorae. This was the basis of the long series of papers by Chapman and Ferraro (1931–1940), who tried to explain magnetic storms and aurorae on the basis of neutral streams.

In 1932 Bartels noted the strong tendency for moderate magnetic storms (presumably not associated with flares) to recur after twenty-seven days (Figure 1.3) and postulated the existence of solar regions active in producing magnetic disturbances, i.e., solar " M-regions." The search for the solar site of the M-regions has been a long one; only recently has there been any hope of approaching a solution (Section 5.6). Many workers have attacked the problem using the method of superimposed epochs, first discussed by Chree and Stagg in 1925. The variation of a chosen geomagnetic index after the central meridian passage of a hypothesized source (such as active regions, sunspots, etc.) can be obtained by superimposing variations of many individual events. A statistically significant peak would indicate a physical correlation, and a time delay would indicate the propagation speed.

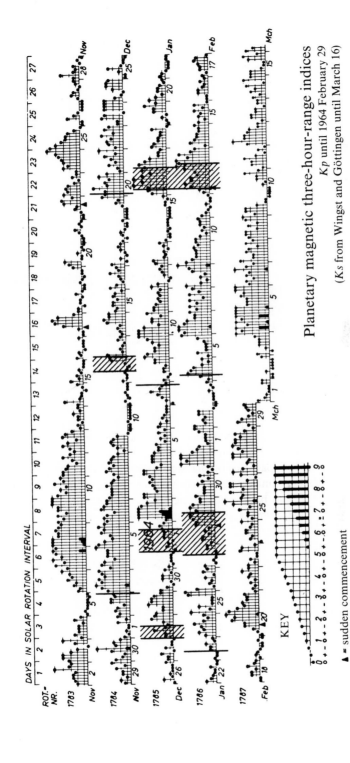

A sample Bartel's musical diagram in the twenty-seven-day format showing clearly the recurrent particle streams or M-regions. Sector boundaries are marked; see Section 5.6. (Courtesy J. M. Wilcox and N. F. Ness.)

Planetary magnetic three-hour-range indices

*Kp* until 1964 February 29

(*Ks* from Wingst and Göttingen until March 16)

KEY

▲ = sudden commencement

The most popular candidates seem to be solar active regions, but, unfortunately, both a minimum and a maximum in geomagnetic activity are found by this method. Thus, two general hypotheses are possible: (1) the active region disrupts a normal, background flow to produce the minimum—the avoidance hypothesis; or (2) the active regions produce the maximum—the hypothesis strongly advocated by Mustel. None of these interpretations is entirely convincing. Note that until about 1960 the auroral-geomagnetic streams were generally thought to be discrete and to move in interplanetary space through a vacuum.

Also note that the auroral theorist concerns himself primarily with the existence of solar corpuscular beams; to him, the solar mechanism for ejection is not that important. The empirical evidence for occasional ejection of solar material is, however, convincing (see Section 2.5); such ejections are associated with flare events, surges, and rising prominences (Figure 1.4). Until the last decade no satisfactory mechanism had been postulated for continuous ejection or emission of solar material. An earlier attempt (1926) was Milne's theory that the mechanism was an outward acceleration produced by the pressure of solar radiation. He thought that because hydrogen is almost completely ionized, the radiation pressure must act through strong absorption lines of other elements with the momentum redistributed through collisions. It has been found, however, that this mechanism is inadequate.

A serious attempt to explain the continuous loss of solar mass evolved from Jean's theory of planetary escape. The solar corona was presumed to have a maxwellian distribution of velocities at a temperature on the order of $10^6$ °K; particles in the tail of the distribution with velocities greater than the solar velocity of escape could be continuously lost from the sun. Exospheric theory approximates the transition from a maxwellian velocity distribution low in the atmosphere to a free-streaming regime high in the atmosphere by the assignment of a critical level or base of the exosphere; it is assumed that collisions dominate and establish an isotropic maxwellian distribution below this level and that above it particles move on individual orbits determined by their injection velocity and by gravitation, i.e., no collisions. The operational definition of the critical level is the height $h_c$ in the atmosphere where an upward-travelling particle with velocity greater than the escape velocity has a probability $1/e$ of escaping without a collision. If the escaping particles are moving through similar particles with an effective radius for collision $a$, we have

$$1 = \int_{h_c}^{\infty} 4\pi a^2 N(h)\, dh \qquad (1.1)$$

For a plane-parallel, isothermal atmosphere, the density variation can be written as

$$N(h) = N(h_c)e^{-(h-h_c)/H} \qquad (1.2)$$

# Solar Events—Sept. 8, 1948

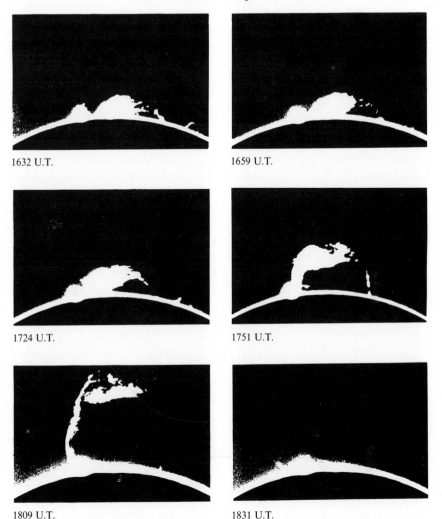

1632 U.T.

1659 U.T.

1724 U.T.

1751 U.T.

1809 U.T.

1831 U.T.

FIGURE 1.4

A sequence of photographs showing a rising prominence. (Courtesy High Altitude Observatory, Boulder, Colorado.)

where

$$H = \frac{kT}{mg} \tag{1.3}$$

is the well-known scale height. Here $T$ is the temperature, $k$ is Boltzmann's constant ($1.4 \times 10^{-16}$ erg/°K), $g$ is the acceleration of gravity, and $m$ is the mass of the particle. If equation (1.2) holds, equation (1.1) integrates to

$$N(h_c) = (4\pi a^2 H)^{-1} \tag{1.4}$$

Physically, equation (1.4) places the critical level where the scale height equals the horizontal mean free path. Coronal densities and a mean collision cross section for protons place the critical level for solar mass loss at $2R_\odot$. The maxwellian velocity distribution is given by

$$f(v) \, dv = 4\pi(m/2\pi kT)^{3/2} v^2 e^{-(mv^2/2kT)} \, dv \tag{1.5}$$

The temperature of the corona is near $2 \times 10^6$ °K (see Section 2.4). The escape velocity differs from the usually quoted gravitational value because any tendency toward charge separation sets up an electric field. Let

$$\beta = \frac{eE}{m_p g_\odot} \tag{1.6}$$

be the ratio of electric to gravitational force on a proton. Then the equation of hydrostatic equilibrium can be written separately for protons and electrons as

$$m_p g_\odot - eE = (1 - \beta)m_p g_\odot = kT_p \frac{d(\log_e N_p)}{dh} \tag{1.7}$$

and

$$m_e g_\odot + eE \approx \beta m_p g_\odot = kT_e \frac{d(\log_e N_e)}{dh} \tag{1.8}$$

The atmosphere is assumed to be isothermal, and the small mass of the electron allows the approximate equality in equation (1.8). There are enough collisions to keep the electron and proton temperatures nearly equal; similarly, an appreciable charge separation cannot occur (because of the strong restoring effects of the electric field that would result), and the proton and electron densities must be nearly equal. Thus, the right-hand sides of equations (1.7) and (1.8) are equal, and this fact implies $\beta \approx 1/2$. Thus, the electric field produces an "effective" proton mass of $m_p/2$, and the escape velocity for protons in a proton-electron mixture becomes

$$v_{esc}(\text{ion}) = \left(\frac{GM_\odot}{r}\right)^{1/2} \tag{1.9}$$

This result is a factor of $2^{1/2}$ lower than the neutral particle value. The situation is complex for other mixtures. The escape velocity for a proton at $2 \times 10^6$ °K from the solar surface is 435 km/sec while the corresponding

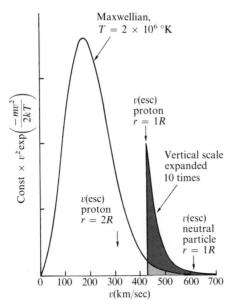

FIGURE 1.5
The maxwellian velocity distribution for $T = 2 \times 10^6 \, ^\circ K$ showing various escape velocities.

neutral particle value is 616 km/sec. Appropriate values are shown in Figure 1.5, together with the maxwellian distribution for $2 \times 10^6 \, ^\circ K$. The particles available for escape are shown in the shaded area. Treatments differ as to the method of replenishing the escaping particles.

Contributions were made by Pikelner and Kiepenheuer, the subject of the coronal mass balance was reviewed by van de Hulst in 1953, and the evaporative-exospheric theory was extended by Chamberlain in 1960. While none of these early evaporative theories proved adequate, it is now possible to construct an exosphere model of the solar wind (Section 3.6). The error of the earlier treatments was in the use of a mean collision cross section in a plasma.

Evidence possibly pertaining to interplanetary gas was also available from study of the zodiacal light and whistlers. As noted above, for some years the zodiacal light had been attributed to the scattering or reflecting of sunlight by interplanetary dust. However, in 1953 Behr and Siedentopf published results concerning the degree of polarization and the brightness of polarized light in the zodiacal light. They found that the polarization was too high to be explained by dust scattering. (Not everyone agreed with their conclusion e.g., Fessenkov.) Hence, Behr and Siedentopf considered the possibility of scattering by interplanetary electrons. They calculated that an electron density near the earth of about $600/cm^3$ would be needed *if all* the polarization was attributed to electrons; occasionally this qualification of their calculations has been ignored.

The electron-scattered content of the zodiacal light was estimated by Blackwell and Ingham in 1961 from the widths of Fraunhofer lines in the zodiacal light spectrum. Because of their small mass, electrons would have large velocities for any reasonable temperature, and solar absorption lines would be broadened in the scattered spectrum. No definite broadening was observed, however, and Blackwell and Ingham found that the electron density near earth could not exceed about $100/cm^3$. Thus, they concluded, dust particles must be responsible for the zodiacal light.

Whistling atmospherics, or whistlers, provide information on the environment a few earth radii from the earth. They are electromagnetic waves with audio wavelengths produced by lightning strokes and channeled along the earth's magnetic field lines between the northern and southern hemispheres. The frequency distribution in successive whistlers is a function of the magnetic field and electron density along the path. If the field is assumed, the density can be obtained, and electron densities of a few hundred/$cm^3$ at distances of 3 to 4 earth radii result. We now know that such distances refer to an environment well within the earth's magnetosphere (Section 6.3), but in the late 1950s such numbers agreed with other determinations.

Finally, evidence concerning solar particle emission has been available from studies of cosmic ray intensities and their time variation. So-called "normal" cosmic rays are of galactic origin; solar cosmic rays are produced from time to time in flare events (and perhaps by other mechanisms also). The galactic cosmic rays appear to be modulated in several different ways. "Forbush decreases" is the term for sudden decreases in the cosmic ray intensity; these usually occur a day after some large flares. This phenomenon could be explained if the earth is suddenly enveloped by a magnetized plasma cloud ejected from the sun as a consequence of the flare event. The magnetic field in the cloud would prevent the ambient cosmic rays from reaching the earth. Forbush also showed that the intensity of galactic cosmic rays varied inversely with the sunspot cycle. This could be explained if the sun emitted larger amounts of magnetized material near solar maximum than at solar minimum (Section 6.7).

## 1.3   The Modern Era

Many important investigations led directly to our modern ideas concerning the solar wind; thus, the classification in "the modern era" is by importance rather than by the date the work was performed. Hoffmeister published the results of an extensive study of the orientations of ionic comet tails in 1943. The importance of this work is often not fully appreciated even though it is among the first quantitative (and pre-discovery!) studies of the solar wind; his observations and results are of value to this day (Section 4.1). One of his graphs contains a plot of the tangent (tan $\varepsilon$) of the angle by which the tail axis lags (in the sense of the comet's motion) behind the prolonged radius vector against the velocity ($V_\perp$) of the comet perpendicular to the radius vector; the points fall approximately on a straight line. The explanation is entirely analogous to Bradley's explanation of stellar aberration: that it is determined by the vector velocity of the earth and the velocity of light. The orientation is determined by the relative vector velocities of the radially expanding solar plasma and the comet in its orbit; the line indicating the relation between tan $\varepsilon$ and $V_\perp$ would have a slope of approximately $w^{-1}$ where $w$ is the solar wind speed. The rough, unweighted value of the solar wind speed determined by Hoffmeister's graph is $\approx 400$ km/sec. A detailed analysis of Hoffmeister's data (see Section 4.1) gives an average solar wind speed of $474 \pm 21$ (p.e.) km/sec. This is in excellent agreement with contemporary values. Hoffmeister's discussion shows remarkable insight into the problem. He speaks of electromagnetic forces due to clouds of charged particles and actually uses the term *Korpuskularstrahlen* (corpuscular radiation). It is curious, nevertheless, that Hoffmeister did not calculate the velocity of the solar wind from his material.

Biermann, from 1951 to 1957, published a series of important papers based on studies of ionic comet tails which led to his postulate of a continuous solar particle emission or solar corpuscular radiation. Note the contrast between this postulate and the generally accepted auroral-geomagnetic view of discrete streams of material moving through a vacuum. Biermann's work largely treated the problem of the accelerations of the knots or clouds of gas (mainly $CO^+$) in the ionic tails. Successive photographs of a comet tail showing a cloud moving along the tail were measured to determine the velocity and acceleration of the cloud; repulsive accelerations about $10^2$ times the local solar gravity were repeatedly found, as well as occasional larger values. The results are usually quoted in the literature in terms of the quantity $(1 - \mu)$, which is the extra radial force of unspecified origin that varies as $r^{-2}$ in units of the gravitational attraction. Thus, the parameter $\mu$ is a direct measure of the effective gravity on the tail, and $(1 - \mu) \approx 10^2$ indicates a force $10^2$ times gravity and in the opposite direction. These were difficult to explain as the result of radiation pressure, as Wurm had noted in 1943. The radiation pressure is simply the available solar photon momentum times the effective area of the atom or molecule, viz.,

$$F_r = \frac{[\pi F_\nu]}{c} \frac{\pi e^2}{m_e c} f \qquad (1.10)$$

where $f$ is the oscillator strength ($f$ value). Inserting the Planck law for the solar radiation flux $[\pi F_\nu]$ yields

$$F_r = \frac{2\pi^2 h e^2}{m_e c \lambda^3} \left(\frac{R_\odot}{r}\right)^2 \frac{f}{(e^{hc/\lambda kT} - 1)} \tag{1.11}$$

Here, besides standard atomic symbols, $R_\odot$ is the solar radius, $r$ is the helio-centric distance, $T$ is the temperature which characterizes the solar radiation field, and $\lambda$ is the wavelength of the transition responsible for the radiation pressure. Some years ago, the $f$ values for the principal transitions of $CO^+$ in the visual and photographic part of the spectrum were not well known. This is no longer the case, and radiation pressure fails by several orders of magni-tude to explain the apparent accelerations found in ionized comet tails.

Biermann attributed the accelerations to momentum transfer to the tail knots from the corpuscular radiation; production of the $CO^+$ ions from the neutral molecule could result from charge exchange with the proton compo-nent. He developed a simple model for the interaction between a proton-electron stream and an ion-electron gas and obtained an expression for the acceleration on the ions as

$$\frac{dv_i}{dt} \approx \frac{e^2 N_e v_e m_e}{\sigma m_i} = 10^{-4.3} \frac{m_e}{m_i} N_e v_e \tag{1.12}$$

Here $\sigma$ is the electrical conductivity which has been evaluated approximately, $m_e$ and $m_i$ are the masses of the electrons and ions, respectively, $N_e$ is the electron density in the beam, and $v_e$ is the bulk velocity of the electrons in the beam. For an assumed velocity of 1000 km/sec (based on geomagnetic delay times) for the corpuscular radiation, equation (1.12) requires a density of about 600 electrons/cm$^3$ to produce an acceleration of $10^2$ cm/sec$^2$, which is close to a $(1 - \mu)$ of $10^2$, because at the earth the solar gravity is approximately 1 cm/sec$^2$ ($=0.59$). This density was entirely compatible with other evidence available at that time, but it is now known to be much too high. However, inclusion of the interplanetary magnetic field would certainly enhance the coupling.

Biermann also explicitly noted the strong support for his hypothesis in Hoffmeister's work and in correlations with geomagnetic indices. By 1957, he had proposed that the sun emits "corpuscular radiation" in essentially all directions at essentially all times. This work was not done entirely in a vacuum, for in 1949, Bartels, in describing the geomagnetic index $Kp$ wrote that it was "designed to measure the varying intensity of solar particle radiation by its geomagnetic effects."

It is essential to note that the existence of the solar particle flux was sug-gested by several lines of evidence, as some of them are in doubt. The mecha-nism by which the solar wind influences and changes $Kp$ is not generally agreed upon. The interpretation of the accelerations of tail knots is questionable; the standard derivations assume that real mass motions are involved, but an alternate interpretation in terms of a wave or excitation phenomenon has been suggested. Of all the early evidence, the aberration effect as reflected in the tail orientations has been the most durable.

The stream of evidence for solar mass ejection is interrupted in our historical narrative by one very significant investigation. In 1957 Chapman considered a "new" model of the corona. It was assumed that the corona and its distant extensions were static and that energy was transferred only by conduction; all other sources of energy transport such as radiative loss were assumed to be negligible. The temperature distribution is determined by conservation of the conductive flux or

$$\mathbf{V} \cdot (\kappa \mathbf{V} T) = 0 \tag{1.13}$$

where $T$ is the temperature, and the coefficient of heat conduction for an ionized gas is

$$\kappa = \kappa_0 T^{5/2} \text{ erg/cm-sec-}°\text{K} \tag{1.14}$$

where

$$\kappa_0 \approx 5 \times 10^{-7} \text{ (c.g.s.)} \tag{1.15}$$

Equation (1.13) is equivalent to

$$\kappa_0 T^{5/2} \frac{dT}{dr} r^2 = \text{const} \tag{1.16}$$

which integrates to

$$\frac{T}{T_0} = \left(\frac{r_0}{r}\right)^{2/7} \tag{1.17}$$

where the subscript zero refers to a reference level usually taken to be the corona. For a coronal temperature of $2 \times 10^6 °\text{K}$, the temperature at earth is $4.4 \times 10^5 °\text{K}$. This means (according to Chapman's model) that the earth is enveloped by a hot plasma. The belief that an additional heat source was necessary for the earth's upper atmosphere accounted for some of the support this model received at that time, but at present, this is not considered a problem. The density at the earth was determined by inserting the derived temperature distribution into the equation of hydrostatic equilibrium, and then densities of 300–400 electrons/cm$^3$ were obtained; again, such values were entirely compatible with most other contemporary values. In fact, the whistler data, the zodiacal light data, the comet observations, and Chapman's model all "agreed" in assigning a value for the electron density near earth within a factor of 2 of 500/cm$^3$; Parker's early values would also "agree" with this.

Chapman's model made two fundamental points: (1) it emphasized the importance of conduction in the maintenance of the corona and its extension; and (2) it regarded the gaseous medium near the earth as simply an extended part of the corona. This was quite important, considering the early discussions of the zodiacal light, the solar H II region, etc., in which no physical connection was assumed between the sun and the distant material. Nevertheless, there were difficulties with Chapman's model which Parker pointed out in 1958.

Parker proposed a hydrodynamic model in which material flowed out of the sun as a natural consequence of the million degree temperature of the corona. He was fully aware of Biermann's work on the "corpuscular radiation" and carefully examined Chapman's model. He found that the density on the static model actually went through a minimum and then increased. The pressure remained finite as $r \to \infty$ and was much larger than any reasonable interstellar pressure (gas or magnetic) which might be invoked for balance. Inclusion of loss mechanisms (e.g., radiation) did not provide a satisfactory resolution of the dilemma. Parker concluded that unless the gas were arbitrarily held static, the corona must expand. His hydrodynamic treatment contained a transonic solution which reproduced the velocities required by Biermann's comet work and had a zero pressure as $r \to \infty$. Parker christened this phenomenon the "solar wind," by which name it has been known ever since.

Parker's first models had an interior region with a constant temperature and an adiabatic temperature variation outside. The early models had a density of about $10^3$ electrons/cm$^3$, velocity of about $10^3$ km/sec, and a temperature of about $10^5$ °K at the earth. These values were not in conflict with evidence available at that time, although one could question the high velocities in the corona implied by the "high" flux at the earth and the equation of continuity. There was also considerable discussion concerning the applicability and stability of the transonic solution. The assumptions implicit in Parker's model are discussed in Section 3.1. Parker also calculated that the solar wind and solar rotation would draw out the solar magnetic field lines into an Archimedes spiral.

Some of the objections to the concept of a solar wind as proposed by Parker were reasonable, and some were not. It is fair to say that the concept was not widely accepted; in fact, for some time the work was involved in controversy. However, the idea of a solar wind was sufficiently attractive to warrant measurement by space probes. This requires a spacecraft sent to a distance of at least 15–20 $R_\oplus$ on the earth-sun line and farther in other directions in order to escape the influence of the earth's magnetosphere (see Section 6.3).

The first *in situ* measurements of the solar wind were made in 1959 by a group of Soviet scientists working under Gringauz with instruments on board Lunik III and Venus I. The data indicated a directed stream of protons with a flux in the range $10^8$–$10^9$/cm$^2$-sec; no evidence for a stationary component of the interplanetary plasma was found. The measurements made on these two flights referred, however, to a rather limited period of time. The first solar wind measurements made by American experimenters were carried out by an M.I.T. group on board Explorer 10 in 1961. Average properties of the medium were as follows: a flux in the range 1–2 × $10^8$/cm$^2$-sec, an average velocity of 280 km/sec, an average proton density of 3–8/cm$^3$, and a mean proton temperature (or equivalent velocity dispersion) of $10^5$–$10^6$ °K. The flux values were consistent with the earlier Soviet determinations, and all values were qualitatively consistent with subsequent results. Unfortunately, Explorer 10 was very close to the boundary of the magnetosphere, and there is no assurance that the results referred to the solar wind "undisturbed" by the presence of the earth.

All reasonable doubt concerning the existence of an essentially continuous solar wind was removed in 1962 by measurements made on board the Venus probe, Mariner 2. Approximately three continuous months of data were obtained. The results as reported by Neugebauer and Snyder were remarkable indeed. The principal features were: (1) a detectable solar wind was present at all times; (2) the average solar wind speed was 500 km/sec; (3) the speed of the solar wind varied between 300 and 860 km/sec, and the velocities were correlated with geomagnetic activity; (4) the average proton density was about 5/cm$^3$; (5) several streams of high velocity plasma were found to reoccur at twenty-seven-day intervals (recall the M-regions discussed above); and (6) alpha particles were discovered to be present in the solar wind. The values cited here are the result of extensive analysis and may, in some cases, differ from the preliminary results.

In the years following 1962 and the Mariner 2 results, scientists have been very active in all phases of solar wind research. The interplanetary magnetic field has been measured and the configuration found (on the average) to fit Parker's description of an Archimedes spiral. Considerable fine structure has been found both in the magnetic field and in the plasma motions. The magnetic field has been found to be closely related to the photospheric field and arranged in segments of predominantly one polarity, which were called sectors by Wilcox and Ness. The field couples the solar wind plasma to the sun and results in a loss of solar angular momentum sufficient to be important in the structure and evolution of the sun. Finally, elements heavier than hydrogen and helium have been detected in the solar wind; the observation of these elements provides valuable clues to the physics of the corona and the origin of the solar wind. All of these phenomena exist because the sun maintains a 2 × $10^6$ °K corona as its outermost atmosphere. In the next chapter, we summarize the relevant features of solar physics.

# Bibliographical Notes: Chapter 1

*General references*

Two books are entirely devoted to the subject of the solar wind. The one presenting the theoretical viewpoint is:

1. Parker, E. N.: *Interplanetary Dynamical Processes*, Interscience Publishers, New York (1963).

The other contains the proceedings of a conference held in April 1964 at the California Institute of Technology:

2. Mackin, R. J., and M. Neugebauer, Eds.: *The Solar Wind*, Pergamon Press, New York (1966).

Current experimental and theoretical results on the solar wind appear in *The Journal of Geophysical Research, Solar Physics*, the volumes of *Space Research* (Proceedings of the yearly COSPAR meetings), and the *Astrophysical Journal*. The *Astronomische Jahresbericht* now contains the heading "corona, solar wind" (*Korona, Sonnenwind*).

A summary of the solar wind was given several years ago in Chapter Seven of the text:

3. Brandt, J. C., and P. W. Hodge, *Solar System Astrophysics*, McGraw-Hill Book Co., New York (1964).

*Section* 1.1

Some review articles that discuss the prehistorical situation are:

4. Chamberlain, J. W.: *Advances Geophys.* **4**, 109 (1958).
5. Dessler, A. J.: *Revs. Geophys.*, **5**, 1 (1967).
See also Reference 1.

The first ideas concerning the zodiacal light are contained in:

6. Cassini, J. D.: *Mem. Acad. Sci.*, Paris, **8**, 121 (1666–1699).

An early work on the forms of comet tails is:

7. Bessel, F. W.: *A.N.*, **13**, 185 (1836).

Many references to early work on comets are contained in:

8. Marsden, B. G.: *A. J.*, **73**, 367 (1968).

The first recorded observation of a solar flare is discussed in:

9. Carrington, R. C.: *Monthly Notices Roy. Astron. Soc.*, **20**, 13 (1859). See also the paper by R. Hodgson following Carrington's.

Some early ideas concerning solar particle emission are found in:

10. FitzGerald, G. F.: *The Electrician*, **30**, 481 (1892).

11. Lodge, O.: *The Electrician*, **46**, 249 (1900).
12. FitzGerald, G. F.: *The Electrician*, **46**, 287 (1900).

*Section* 1.2

The references for the early auroral work are contained in References 4 and 5.

The solar M-regions are hypothesized in:
13. Bartels, J.: *Terr. Mag. Atm. Elect.*, **37**, 1 (1932).

References and results concerning the method of superimposed epochs are found in:
14. Mustel, E. R.: in Annals of the IQSY, **4**, *Solar-Terrestrial Physics: Solar Aspects*, M.I.T. Press (1969), p. 67.

The historical reference is:
15. Chree, C., and J. M. Stagg: *Phil. Trans. Roy. Soc.* (London), **A227**, 21 (1927).

A mechanism for solar mass loss based on radiation is contained in:
16. Milne, E. A.: *Monthly Notices Roy. Astron. Soc.*, **86**, 459 (1926).

Also of interest is:
17. Ellison, M. A.: *The Sun and Its Influence*, Routledge & Kegan Paul, London (1959).

Exospheric theory as applied to the solar corona is reviewed in:
18. van de Hulst, H. C.: in *The Sun*, G. P. Kuiper, Ed., University of Chicago Press, Chicago (1953), p. 306.
19. Chamberlain, J. W.: *Ap. J.*, **131**, 45 (1960). (Note a numerical correction to this work in Reference 7 of Chapter Three.)

The effective gravity in ionic mixtures is discussed in:
20. Parker, E. N.: in *The Solar Corona*, J. W. Evans, Ed., Academic Press, New York (1963), p. 11.

The pertinent zodiacal light work is contained in:
21. Behr, A., and H. Siedentopf: *Zs. f. Astrophys.*, **32**, 19 (1953).
22. Blackwell, D. E., and M. F. Ingham, *Monthly Notices Roy. Astron. Soc.*, **122**, 113 (and following papers) (1964).
23. Beggs, D. W., and D. E. Blackwell, D. W. Dewhirst, and R. D. Wolstencroft: *Monthly Notices Roy. Astron. Soc.* **127**, 319 (and following paper) (1964).

Elementary reviews and additional references concerning the zodiacal light, whistlers, and the cosmic ray events are found in Reference 3; see also Reference 1.

*Section* 1.3

The early fundamental work on ionic comet tails is found in:
24. Hoffmeister, C.: *Zs. f. Astrophys.*, **23**, 265 (1943).
25. Biermann, L.: *Zs. f. Astrophys.*, **29**, 274 (1951).
26. Biermann, L.: *Mem. Soc. Sci. Liège*, 4th ser., **13**, 251 (1953).
27. Biermann, L.: *Observatory*, **77**, 109 (1957).

See also the following two reviews, which cover the acceleration of tail knots, and the references given there:

28. Biermann, L., and Rh. Lüst: in *The Moon, Meteorites and Comets*, B. M. Middlehurst and G. P. Kuiper, Eds., University of Chicago Press, Chicago (1963), p. 618.
29. Wurm, K.: in *The Moon, Meteorites and Comets*, B. M. Middlehurst and G. P. Kuiper, Eds., University of Chicago Press, Chicago (1963), p. 573.

The description of the interrelation between the solar particles and geomagnetic effects is found in:

30. Bartels, J.: *J. Geophys. Res.*, **54**, 296 (1949).

The static conductive model of the interplanetary plasma is given in:

31. Chapman, S.: *Smithsonian Contrib. Astrophys.*, **2**, 1 (1957).
32. Chapman, S.: *Proc. Roy. Soc.*, A, **253**, 450 (1959).
33. Chapman, S.: in *Space Astrophysics*, W. Liller, Ed., McGraw-Hill Book Co., New York (1961), p. 133.

Compare the conceptual approach with:

34. Elsässer, H.: *Mitt. Astron. Ges.*, **1957** II, 61 (1957).

The basic ideas concerning the hydrodynamic expansion of the corona are given in:

35. Parker, E. N.: *Ap. J.*, **128**, 664 (1958).

Subsequent early theoretical work is summarized in Reference 1.

The spacecraft measurements of solar wind plasma and magnetic field are presented in Chapter Five of this work; see the references given there.

# 2

# *A Summary of*
# *Solar Physics*

This chapter, as its title implies, is a review of basic solar physics, with emphasis on: (1) the generation and transport of the mechanical energy believed responsible for the existence and heating of the solar corona and (2) the physical properties of the corona and the overall aspects of solar activity.

## 2.1  Solar Structure

The mass ($M_\odot = 1.99 \times 10^{33}$g) and chemical composition of the sun determine the solar radius ($R_\odot = 6.96 \times 10^{10}$cm) and the total solar luminosity ($L_\odot = 3.9 \times 10^{33}$ ergs/sec); any successful model of the solar interior must reproduce these directly observable quantities. The chemical composition of the photosphere is inferred from studies of the Fraunhofer lines in the solar spectrum. The photospheric composition is generally assumed to be indicative of the composition of the initial sun, although the composition of the deep interior has been altered by the nuclear reactions responsible for the solar luminosity.

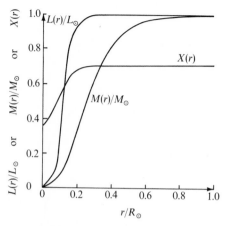

FIGURE 2.1

The variation of luminosity, mass, and the fraction of hydrogen (X) on the basis of Sears' model of the sun. Note the central depletion of hydrogen. The function L(r) and M(r) denote the luminosity and mass, respectively, interior to r.

The solar luminosity is produced in the sun's central regions primarily via the proton-proton chain, although the carbon cycle does make a small contribution. The energy of the proton-proton chain is transferred throughout the deep interior by radiation, the opacity being due primarily to bound-free transitions. Most (but not all) atoms are completely ionized in the deep solar interior, and the perfect gas law is an accurate equation of state.

The equations needed to construct a model of the solar interior are: (1) hydrostatic equilibrium; (2) mass distribution; (3) energy balance; (4) energy transfer; (5) equation of state (the perfect gas law; radiation pressure is unimportant in the sun); (6) the opacity law; and (7) the rates of energy generation. These equations and the appropriate boundary conditions determine the properties of an initial solar model. To obtain a model of the present sun, it must be evolved to an age of $4.5 \times 10^9$ years. Such a model is shown in Figures 2.1 and 2.2. The detailed construction of such a solar model is carried out by use of numerical methods on a high-speed electronic computer.

The solar magnetic field and solar rotation have usually been ignored in discussions of the solar interior. This was because there was very little information from surface observations about the strength and configuration of the magnetic field in the interior and there was also no conclusive evidence for departure from a spherical shape to cite as an effect of rotation. However,

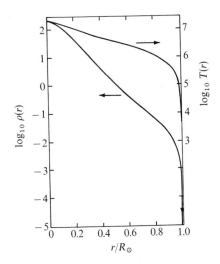

FIGURE 2.2
The variation of density and temperature
on the basis of Sears' solar model.

Dicke and Goldenberg have reported a measurement of the apparent solar
oblateness and give

$$\varepsilon = \frac{R(\text{eq}) - R(\text{pole})}{R_\odot} = (5.0 \pm 0.7) \times 10^{-5} \tag{2.1}$$

The observations are difficult, but the reality of the results is supported by the
fact that the apparent oblateness changed as the sun's pole changed orientation
with respect to the observing coordinate system. The relationship of the
apparent to the dynamical oblateness is not entirely straightforward; however,
the figure quoted in equation (2.1) could apply to a dynamical oblateness and,
if so, implies a solar core that rotates in about one day. Such a core would
supply angular momentum to the solar atmosphere, and this steady-state
situation would require an angular momentum sink, which the solar wind
might provide (see Sections 7.2 and 7.3). The steady-state situation could also
be related to the solar differential rotation in light of the fact that the photospheric rotation period increases with latitude.

Another possible problem in the solar interior concerns the rate of neutrino
emission. Nuclear reactions in the solar interior produce neutrinos, which
escape freely from the solar interior providing a unique opportunity to
" observe " it. Predictions based on models of the solar interior and knowledge
of the nuclear processes have been made, but to date these have *not* been
confirmed. The lack of such confirmation may indicate a fundamental uncertainty in our knowledge of the solar interior.

Among the equations used to determine the structure of the solar interior was the energy transfer equation. Energy transfer proceeds by radiation throughout most of the solar interior, but convective transport can dominate near the surface. In model construction, each point must be tested for convective instability and the appropriate equation then used. In the sun, a convective zone exists from $r/R_\odot = 0.86$ to just below the photosphere. The test is made with the condition for convective instability, the Schwarzschild criterion, which is

$$\left|\frac{dT}{dr}\right|_{str} > \left|\frac{dT}{dr}\right|_{ad} \qquad (2.2)$$

where the subscripts str and ad refer to the structural and adiabatic temperature gradients, respectively. Consider a bubble undergoing a small, adiabatic displacement upward and remaining in pressure equilibrium with the surrounding material. For the condition given by equation (2.2), the bubble is hotter than the surrounding gas and, hence, at a lower density then the surroundings. The buoyancy force acts on the bubble to continue the upward motion, and the situation is unstable. The Schwarzschild criterion is a necessary but not sufficient condition for convection. The Rayleigh number (discussed in Section 2.2) must also exceed a critical value. In the present solar model, the Schwarzschild criterion indicates convection from $r/R_\odot \simeq 0.86$ to just below the photosphere.

## 2.2   Hydrogen Convection Zone

The physical condition for convection is a high structural temperature gradient, which is usually determined by the relative efficiency of energy transfer by radiation. Hence, a high radiative temperature gradient implies a small mean free path for photons or, equivalently, a relatively high opacity. As shown by Wildt in 1939, the opacity in the solar photosphere is due primarily to the negative hydrogen ion, $H^-$. The abundance of $H^-$ in the photosphere is accurately given by the Saha equation which has the form

$$\frac{N(H)N_e}{N(H^-)} = f(T) \approx \text{const} \times T^{3/2}\, e^{-I/kT} \qquad (2.3)$$

where I is the ionization energy for $H^-$; the binding energy is 0.75 ev for the outer electron, which places the ionization limit at about 16,500 Å. The opacity per unit mass is determined by the ratio of $N(H^-)$ to $N(H)$ which is proportional to $N_e/f(T)$. Hydrogen is largely neutral in the photosphere, and the electrons are provided by metals with low ionization potential. These metals are Mg, Si, Fe, Ca, Al, and Na; the aggregate of these metals is approximately $10^{-4}$ the abundance of hydrogen. Thus, at (say) 5,000 °K, the abundance of $H^-$ is limited by the number of electrons available from metals. At $\approx 10^4$ °K, hydrogen is approximately half ionized, and the number of available electrons has increased by orders of magnitude. Since $f(T)$ in equation (2.3) is slowly varying within the range under consideration, the relative abundance of $H^-$ is dramatically increased in a short distance of 200–300 km. This increase determines the beginning of convective instability beneath the photosphere. At deeper levels, there are plenty of electrons but no neutral hydrogen, which accounts for the lack of importance of $H^-$ in the determination of the opacity and convection ceases.

The details of the formation and extent of the hydrogen convection zone (HCZ) are not well understood; nevertheless, the basic physical nature is qualitatively clear. A regime of rising hot elements (bubbles) and falling cool elements is set up. The nature of the regime is determined by the dimensionless Rayleigh number $\Lambda$ which can be considered as the ratio of impulses on a bubble resulting from buoyancy and viscous forces. If the viscous forces dominate the buoyancy forces, $\Lambda$ is small and no convection occurs; if the Rayleigh number exceeds a critical value $\Lambda_0$, convection takes place. As $\Lambda$ is increased beyond the critical value, the convection occurs in a regular cell structure resembling the Bénard cells of classical laboratory studies. As $\Lambda$ is further increased, the convection first becomes nonstationary and then random or chaotic.

Available theory and pertinent parameters predict that random convection exists in the solar HCZ. This does not appear to be the case in Figure 2.3, for example, where the solar granulation probably corresponds to the overshoot of convective elements which penetrate into the photosphere. (Mean "cell" sizes of about 700 km have been derived.) Actually, seeing effects may seriously influence estimates calculated from ground-based observations. The granulation pattern consists of hot rising elements surrounded by cold falling elements; lifetimes of about five minutes are found for individual elements. Such relatively long lifetimes are not theoretically expected (although they are consistent with typical velocities of $\sim 1$ km/sec and a mixing length of $\sim 300$ km) with the result that the use of the atomic viscosity in the calculation of the Rayleigh number becomes suspect. Turbulent viscosity is larger than atomic viscosity and could lower the Rayleigh number to within the correct range. In fact, the Rayleigh number could be lowered to such an extent that convection would not be expected. Both results are contrary to observation, but the relative discrepancy is much less for the turbulent viscosity. Physically, this means a sharp increase in the ability of "viscous" forces to control and stabilize the convective motions.

The basically turbulent nature of the convective motions is crucial in the discussion of the origin of the mechanical noise that is thought to heat the chromosphere and corona. The flow can be described by the Reynolds

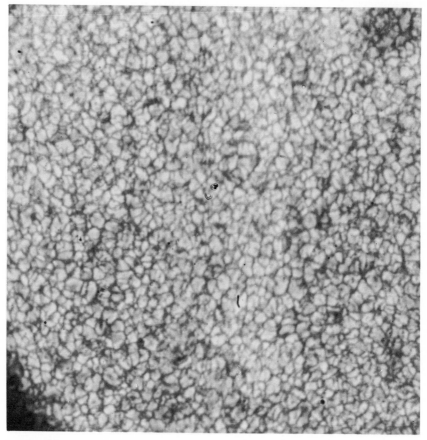

FIGURE 2.3

Solar granulation photograph taken with a balloon-borne telescope. (Courtesy Project Stratoscope of Princeton University, sponsored by the Office of Naval Research, the National Science Foundation, and the National Aeronautics and Space Administration.)

number

$$\mathrm{Re} = \frac{vl}{v} \tag{2.4}$$

where $v$ is the velocity of the convective element, $l$ is a characteristic length, and $v$ is the coefficient of kinematic viscosity. The kinematic viscosity is approximately $(1/3)\lambda v_{\mathrm{th}}$ where $\lambda$ is the mean free path and $v_{\mathrm{th}}$ is the thermal velocity $(= [8kT/\pi \mu m_H]^{1/2})$. Physically, the Reynolds number is the ratio of the force required to produce the acceleration $(\rho v^2/l)$ to the viscous force $(\rho v v/l^2)$. If the viscous forces are large enough, they can keep the flow laminar.

When the Reynolds number exceeds a critical value, which is usually $Re \approx 10^3$, the flow becomes turbulent. In the subphotospheric layers, $v = 10^3$, $l$ is of the same order as the scale height or $10^7$ cm, and the convective motions are in the range 1 to 3 km/sec. Thus, the Reynolds number in the HCZ is $\sim 10^9$, and turbulent flow is definitely expected. This turbulent flow is presumed to be the ultimate source of the energy which maintains the chromosphere and corona.

Besides the motions of $\sim 2$ km/sec associated with the solar granulation, other motion patterns are known to exist in the photosphere. One is supergranulation, which consists of a network of cells with a typical dimension $\approx 30,000$ km and a very approximate lifetime of about one day. The flow is from cell center to rim with a velocity of about 0.4 km/sec; vertical velocities involved are somewhat smaller. Another motion pattern is the oscillatory field with a typical size somewhat larger than the ordinary granules. This vertical motion has an r.m.s. amplitude of about 0.4 km/sec and an average period of about 300 sec. These motions presumably have their ultimate origin in the HCZ and could be the source of waves propagating into the corona and the solar wind.

The supergranulation motion has the effect of convecting the frozen-in, photospheric magnetic fields to the cell boundaries. The supergranulation boundaries show a close correspondence with the chromospheric network (see Figure 2.16). The spicules (thin, brilliant elements from which the chromosphere is composed at heights of 5,000 km above the limb and higher) originate in the supergranulation boundaries. This pattern of fine structure may extend into the solar wind (see discussion in Section 5.6).

## 2.3   Origin of the Chromosphere and Corona

The temperature structure of the solar atmosphere (photosphere, chromosphere, and corona) is shown in Figure 2.4; these are the results of traditional solar studies and serve to illustrate the problem of coronal maintenance. Consider an atmosphere in radiative equilibrium—that is, each volume element absorbs and emits exactly the same amount of radiant energy; thus, no other mechanisms for energy transport are important. If one considers a gray atmosphere (meaning that the absorption coefficient is independent of frequency), the temperature distribution can be determined in terms of the effective temperature. The temperature decreases monotonically with increasing height in the atmosphere, and the "boundary" temperature $T_0$ is given in terms of the effective temperature $T_e$ by $T_0^4 = (3^{1/2}/4)T_e^4$; numerically, this relation is $T_0 = 0.81\, T_e$. The solar effective temperature is about 5,750 °K, and hence a boundary temperature of 4,600–4,700 °K would be expected. While the assumptions involved in this discussion seem idealistic, the basic

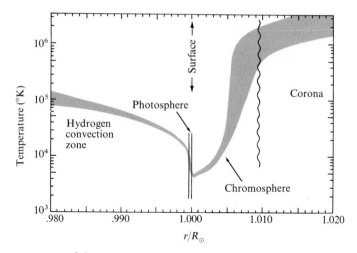

FIGURE 2.4

The temperature profile in the upper hydrogen convective zone,
photosphere, chromosphere, and lower corona.

result is generally accepted. The existence of a temperature minimum and
maximum temperatures of $10^5$ and $10^6$ °K in the atmosphere clearly indicates
a nonradiative source of energy. In the 1940s, Alfvén, Biermann, and
Schwarzschild suggested that the mechanical noise generated in the HCZ
could heat the outer atmosphere.

The picture suggested by them is basically as follows. The rate of generation
of acoustical noise per unit volume is roughly known from theory and varies
as high power of the velocity. The total noise production from the entire
HCZ can be calculated from a model. Since the velocities in the HCZ show a
peaked distribution and because of the steep velocity dependence of the noise
generation and the temperature dependence, most of the energy is generated in
a thin layer with a thickness on the order of 100 km. Thus, we have the
generation of a flux of acoustic waves which travel upward (and downward) at
the local speed of sound. The upward flux is

$$F_n = \tfrac{1}{2}\rho(\delta v)^2 v_s \tag{2.5}$$

where $\rho$ is the density, $\delta v$ is the amplitude of the material velocity, $v_s = (\gamma P/\rho)^{1/2}$ is the speed of sound, $\gamma$ is the ratio of specific heats, and $P$ is the
pressure. Note that the velocity of sound is essentially proportional to the
square root of the temperature.

If the properties of these waves and the atmosphere are such that no dissipation takes place, then $F_n$ is a constant. Under such conditions, consider how
the quantities in equation (2.5) vary as these waves propagate upward in the
solar atmosphere. The velocity of sound would vary by only a factor of about

ten if the temperature varied from $10^4$ to $10^6$ °K; the variation through the upper HCZ, photosphere, and lower chromosphere is much less. On the other hand, the density drops by about five powers of ten between the region of maximum noise generation and the transition zone between the chromosphere and corona; the transition zone is apparently rather turbulent, and it could be the area where mechanical energy deposition begins to be important. The only way the noise flux can remain constant is for the velocity amplitude to increase by a factor of about one hundred from approximately 0.1 km/sec in the region of peak noise generation to approximately 15 km/sec in the transition zone; this last velocity is comparable with the sound speed for temperatures of $10^4$–$10^5$ °K, temperatures which are appropriate to the upper chromosphere and transition zone. From such a situation, shock waves are expected to develop.

This argument is intended to be only qualitative. Empirical evidence exists for energy deposition throughout much of the chromosphere and corona. The formation of shock waves or discontinuities is a fairly general property of simple waves propagating in the direction of decreasing density. Regions of compression travel faster and regions of rarefactions travel slower on the wave profile with respect to the regions of equilibrium density. Thus, the waves eventually steepen into shocks. It is relatively easy to visualize the dissipation of energy in a shock because a discontinuity is involved. The entropy increases through a shock as a result of dissipation; the amount required is determined by the conservation of mass, energy, and momentum. Mechanical energy deposition like mechanical energy generation varies as a high power of the mach number, and energy deposition is expected to begin in the chromosphere and/or transition zone. Theoretical and observational evidence indicates that energy deposition extends to 0.5 to 1.0 $R_{\odot}$ above the surface.

The physical picture sketched above is probably correct in broad outline, but complications abound and the details are hard to fill in. The physical nature of the waves responsible for the heating is still not clear. They may be simple sound waves as outlined above, but the presence of the solar magnetic field complicates the situation. In an incompressible fluid with high electrical conductivity (such as liquid mercury) and a magnetic field B, small disturbances propagate along the magnetic field as transverse Alfvén waves at the Alfvén velocity,

$$V_A = \frac{B}{(4\pi\rho)^{1/2}} \qquad (2.6)$$

where $\rho$ is the density. For the case of disturbances originating in a compressible fluid with a magnetic field, the transverse Alfvén waves interact with the longitudinal sound waves to produce hybrid waves of varying complexity

—the Alfvén mode, the slow mode, and the fast mode. The complications involved, however, are apparent because the wave properties depend on the direction of propagation relative to the magnetic field.

The magnetic field is of additional importance because both mechanical noise generation and deposition are enhanced in the presence of a magnetic field. A detailed calculation would include the computation of the noise generation in the HCZ, a study of the various modes to determine the ones that reach the upper chromosphere or low corona, and the determination of the energy deposition per unit volume in the chromosphere and the corona. Although many efforts have been made, no definitive and generally accepted calculation exists at present. Note that wave energy absorbed (say) in the photosphere has no observable consequences because the flux is small compared with the radiative flux which is strongly coupled to the matter (i.e., the radiation temperature and kinetic temperature are approximately equal), and the additional flux can be easily radiated away.

Other mechanisms for heating the corona through wave deposition have been suggested; these are based on: (1) gravity waves (where gravity is the restoring force) and (2) Landau-damping of ion-acoustic waves generated in the photosphere. While these suggestions should be kept in mind, the details are at present still sketchy. No one seriously considers any longer that the heating of the corona is caused by the accretion of meteoritic material.

Some insight into the problem of coronal maintenance can be obtained from consideration of the coronal energy balance. First, the corona requires some defining; here it is taken to start at $1.1 \, R_\odot$, which means temperatures close to $10^6 \, °K$. If arbitrary boundaries are drawn at $1.1 \, R_\odot$ and $2.0 \, R_\odot$, the primary energy losses can be calculated. The largest amount of loss consists of conduction in approximately equal amounts both out into the interplanetary medium and down into the chromosphere (and contributing to its heating); the conductive losses depend on the temperature and its gradient, which are inferred from the density distribution in the corona (see Section 2.4). Radiative losses make a numerical contribution, but they do not appear to be the major cause of energy loss, as is often assumed. Part of the difference between accountable and unaccountable losses may revolve around the question of the inclusion of the transition zone in the corona. The support of the coronal energy loss requires a source of approximately $10^5 \, \text{ergs/cm}^2\text{-sec}$ in the form of mechanical energy.

An alternate procedure may also be employed. The standard solar wind equations (Section 3.1) are: (1) equation of motion, (2) equation of continuity, and (3) an energy equation, which includes energy deposition. These equations can be solved for the run of energy deposition required to produce the empirically determined run of density in the corona. An entirely reasonable curve results with a peak near $r = 1.3$–$1.4 \, R_\odot$ and which goes to zero before $2.0 \, R_\odot$ is reached. The area under this curve is $\approx 10^5 \, \text{ergs/cm}^2\text{-sec}$. While these

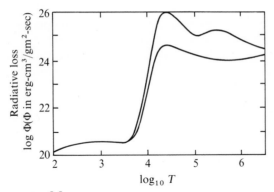

FIGURE 2.5

The variation of the radiative loss with temperature (after
R. Weymann). The upper curve assumes a cosmic
mixture; it also assumes that all de-excitations are
radiative and that no energy gain occurs by photoionization. The lower curve assumes only free-free and
bound-free emission of helium and hydrogen. The
upper curve is probably closer to reality; the quantity
$\rho^2\Phi$ is the radiative loss in ergs/$cm^3$-sec.

results are entirely plausible, they could be due to a breakdown of other
assumptions used in the analysis, such as spherical symmetry. Nonetheless, a
source of mechanical energy is undoubtedly necessary for the maintenance of
the corona.

The role of radiative losses is extremely interesting in reviewing the basic
physics of the corona. The approximate variation of energy loss per unit mass
by radiation is shown in Figure 2.5. Note the maximum near $10^4$ °K where
hydrogen is an efficient radiator; however, at $10^6$ °K, hydrogen is fully
ionized, and radiative losses are relatively inefficient. Thus, the corona exists
because the mechanical energy is dissipated in a region of such low density or
total mass that it is raised to a temperature where radiative losses cannot
dispose of the incoming energy. Thus, the situation becomes one of balance
between input of mechanical energy and outflow by conduction. At large
distances from the sun, the outflowing conductive energy is converted into the
bulk kinetic energy of the solar wind.

## 2.4 Physical Properties
##     of the Corona

The corona is commonly considered a fully ionized plasma at temperatures of
about $10^6$ °K. Here we will review the evidence leading to this conclusion and
present other details of the physical properties of the corona.

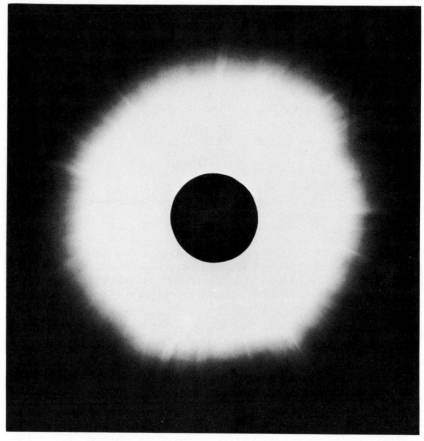

FIGURE 2.6
The corona at solar maximum. (Courtesy G. Van Biesbroeck.)

*Introduction and Basic Structure.* The corona has been traditionally observed at the time of solar eclipse by means primarily of photographs. Some facets of coronal structure are apparent from inspection of the sample coronal pictures shown in Figures 2.6, 2.7, and 7.3. The corona at time of maximum in the sunspot cycle exhibits a fairly close approximation to spherical symmetry and shows fine structure in the form of essentially radial, narrow structures called rays. At intermediate times in the solar cycle, the coronal appearance is asymmetrical and dominated by a small number of coronal features with dimensions $\approx 1 R_\odot$ called fans or helmets. The fans are apparent in the intermediate corona, absent in the corona at solar minimum, and so numerous and widespread at solar maximum that the corona appears rather spherical. Fans are normally associated with a quiescent prominence (Section 2.5), and coronal

FIGURE 2.7
The corona at solar minimum. (Courtesy A. Wallenquist.)

arch systems in the fan surround the prominence. Fine rays are visible in the intermediate corona and very apparent in the minimum corona especially near the poles; these features are the polar rays, or polar plumes, which remind one of a dipole magnetic field. At solar minimum, the very little solar activity that exists is found near the equator; thus, an asymmetrical corona is produced.

A quantitative estimate of this effect can be seen by defining the coronal flattening as

$$\varepsilon = \frac{d_E}{d_p} - 1 \qquad (2.7)$$

where $d_p$ and $d_E$ are the polar and equatorial diameters of a given isophote. Usually the given diameter is averaged with the two diameters $22°5$ to either side to avoid spurious effects due to local irregularities. If the equatorial diameter is taken to be $4\ R_\odot$ (radius of $2\ R_\odot$), the flattening $\varepsilon$ (the Ludendorff index) at solar minimum is usually between 0.25 and 0.30, and the flattening at maximum can approach zero with most values less than 0.10. However, there are individual exceptions to these general variations.

The fine structure in the form of rays is explained by the magnetic field that is suggested by the dipole form of the polar rays. Density irregularities in the corona can be preserved perpendicular to the lines of force because diffusion in this direction is greatly inhibited by the field. No such restriction exists along the lines of force; as a result, the long narrow features that were observed were expected. If the atmosphere is in hydrostatic equilibrium (or a close approximation thereto), the density gradients should be the same both in and out of the rays. This is the case because the pressure gradient is balanced by the gradient of the potential, and no magnetic term appears in the potential. The behavior of equal density gradients is predictable and apparently observed.

The density in a polar plume is about four to five times that of the immediate surroundings. This figure was reached after an analysis of tracings of polar regions on eclipse plates. For the 1900 eclipse, van de Hulst noted that single rays increased the coronal brightness by 10 percent and that the rays were about 7,000 km in diameter. Van de Hulst determined an effective path length of the polar corona and found that an individual ray diameter constituted some one-fortieth of the total. Therefore, the density must be enhanced by a factor of about four to produce a 10 percent increase in brightness. Similar enhancements were deduced from studies conducted during other eclipses. Enhancements of the same general amount over the normal coronal densities are found for streamers and helmets.

Besides the radial, fine structure associated with density fluctuations, there are other structural details often distinguished by enhanced intensity in certain emission lines. First, there are the fans or helmets, as noted earlier, which dominate the gross form of the corona. These are associated with quiescent prominences (Section 2.5); immediately above the prominence is a very dark region which is succeeded by a bright coronal arch system. Kiepenheuer has described the overall appearance of this entity in cross section as resembling the Eiffel Tower. The tip of this structure usually extends one or two solar radii above the limb; the base is usually about one-half a solar radius in width.

Streamers are extensions of the corona over active solar regions. They can be distinguished out to many solar radii and are fairly structureless; their thickness (or cross section) remains fairly constant with increasing radial distance.

Coronal or sporadic condensations are found only over very active or flaring groups. Usually they are found at the top of the active or sunspot prominences and in white light appear with an extent of a few times $10^4$ km. Their density exceeds that of the surrounding medium by one or two orders of magnitude. These condensations have lifetimes of hours, or, at most, days because this is the length of maximum activity associated with a sunspot group (see Section 2.5). Coronal condensations generally show the coronal yellow line of Ca XV at 5,694 Å; this line generally indicates a very high temperature.

Coronal enhancements or permanent condensations have base dimensions comparable to plages (Section 2.5) and in many respects appear to be an extension of the plage activity into the corona. Their density is higher than the background corona but lower than the condensations. The enhancements can last for several solar rotations. During the time of maximum activity associated with a spot group, the innermost dense part of the enhancement may be described as a condensation. The coronal enhancements are often called coronal green patches because they are bright in the coronal green line of Fe XIV at 5,303 Å.

The terminology in studies related to solar activity is often quite confusing even to active workers in the field. Sometimes, for example, the same word is used to refer to two entirely different phenomena.

The coronal light may be divided into three components:

1. The K corona. This is continuous radiation resulting from photospheric radiation that is Thomson-scattered by free electrons in the corona.

2. The F corona (or false corona or inner zodiacal light). This is photospheric radiation diffracted by interplanetary dust. This radiation is not physically connected with the corona but must be accurately determined in order to separate it from the desired K corona.

3. The E or L corona. This is the total light of the coronal emission lines (optical region), such as the coronal green and yellow lines previously mentioned.

The brightness of the various components and other pertinent information are shown in Figure 2.8. The corona is often divided into the inner corona ($r/R_\odot \leq 1.3$), the medium corona ($1.3 \leq r/R_\odot \leq 2.5$), and the outer corona ($r/R_\odot > 2.5$); the latter region merges into the solar wind at large distances.

*Densities.* Coronal electron densities are calculated from photometric determinations of the brightness of the K corona. Often such measurements are presented in the form of isophotes. The K corona is presumed to originate by Thomson-scattering of photospheric radiation with cross section $\sigma = 6.6 \times 10^{-25}$ cm$^2$. The standard treatment assumes an optically thin corona. Thus, the coronal intensity at a particular frequency is simply the integral of the source function per unit volume along the line of sight, viz.,

$$4\pi I_v = R_\odot \int_{-\infty}^{+\infty} g(y)\,dy \tag{2.8}$$

Here $y$ is in units of solar radii. The geometry is shown in Figure 2.9.

In order to continue, one must make some assumptions concerning symmetry which allows for the determination of three dimensional structure

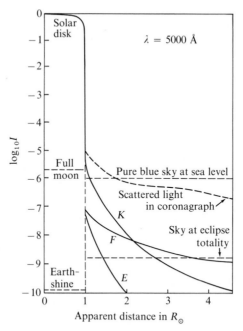

FIGURE 2.8

Schematic of the variation of the K, F, and E components of coronal light and other brightnesses of interest.

from a two-dimensional light distribution. Spherical symmetry is often assumed and is adopted here. The assumption that Thomson-scattering is not isotropic involves very little error and is also used here. The source function per unit volume is then

$$g(r) = \sigma N_e(r) \int_{\substack{\text{solar} \\ \text{disk}}} \frac{I \, d\omega}{4\pi} \tag{2.9}$$

or

$$g(r) = \sigma N_e(r) J(r) \tag{2.10}$$

where $J(r)$ is the mean intensity of radiative transfer theory. It is determined by an integration of the emergent intensity over the solar disk taking account of limb-darkening; $J(r)$ varies approximately as $1/r^2$ and can be taken as known. Equation (2.8) can be written as

$$4\pi I_v = 2R_\odot \int_0^\infty [\sigma N_e(r) J(r)] \, dy \tag{2.11}$$

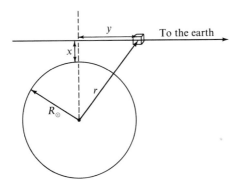

FIGURE 2.9

The geometry and nomenclature for computing the coronal electron density from
observed brightnesses.

It is found by observation that the intensity of the K corona can be accurately expressed by a few terms of a sequence in inverse powers of $r$. Thus, one is
led to consider a source function composed of inverse powers of $r$, or

$$g(r) = \sum_n a_n r^{-(n+1)} \tag{2.12}$$

The quantity $r$ can be expressed in terms of $y$ by considering the geometry as
shown in Figure 2.9, and equation (2.11) can be rewritten for one term of
equation (2.12) as

$$4\pi I_n = 2a_n R_\odot \int_0^\infty \frac{dy}{[(x + R_\odot)^2 + y^2]^{(n+1)/2}} \tag{2.13}$$

$$= \frac{2a_n R_\odot}{(x + R_\odot)^n} \int_0^{\pi/2} \cos^{n-1}\phi \, d\phi$$

Hence, an observed density component that varies as $r^{-n}$ results from a
source function component that varies as $r^{-(n+1)}$. If the intensity varies as a
few terms of such a sequence, then it is a simple matter to determine the
source function producing it; thus, $g(r)$ can be determined directly from
observation. Note that there is no particular physical significance to these
interpolation formulas.

If $g(r)$ is known, then $N_e(r)$ can be determined from equation (2.10) because
$J(r)$ is known. Sample determinations of $N_e(r)$ are shown in Figure 2.10. The
determination of the proton concentration requires the specification of the
helium concentration; a hydrogen to helium ratio of 10:1 gives $N_p = 0.83 \, N_e$.

The assumptions involved in the determination of coronal densities must
also be kept in mind. The corona is assumed to be spherical and homogeneous,

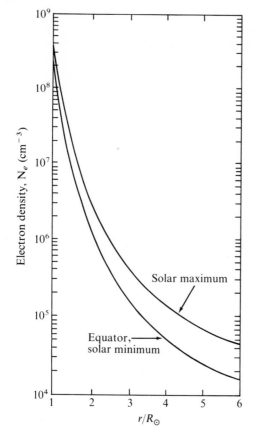

FIGURE 2.10

Some sample determinations of coronal electron density. (After C. W. Allen and H. van de Hulst.)

and the scattering is assumed to be isotropic. While all these assumptions produce some error, the assertion of homogeneity of the corona is probably the most serious. Some measure of the distribution of densities at a given $r$ would be desirable; consider the function

$$\frac{1}{X} = \frac{\langle N_e \rangle^2}{\langle N_e^2 \rangle} \tag{2.14}$$

An entirely homogeneous corona has $1/X = 1$, and this function decreases for a corona with inhomogeneities. A simple interpretation follows from consideration of a sequence of $n$ 1 cm$^3$ boxes, only one of which contains $N_e$ electrons. The quantity $\langle N_e \rangle^2$ becomes $\langle N_e/n \rangle^2$ while $\langle N_e^2 \rangle$ becomes $\langle N_e^2/n \rangle$. Thus, $1/X = 1/n$ and leads to a simple interpretation of $1/X$ as the fraction of space occupied by matter. The densities determined from eclipse observations and the assumption of homogeneity are mean values, $\langle N_e \rangle$. A first step toward introducing the effects of the fine structure is to think of the mean density at a given $r$ as determined by a combination of vacant regions and so-called "model rays," which occupy only $1/X$ of the space but which have a density of $X\langle N_e\rangle$. The function $X$ can be estimated from eclipse photographs (recall the discussion above concerning the determination of the density enhancement in the polar plumes) and the interpretation of radio observations. Allen has compiled the pertinent data, which may be approximately represented in the corona by

$$X = 1.6\left(\frac{r}{R_\odot}\right) - 0.7 \tag{2.15}$$

Thus, at $r = 2R_\odot$ we have $1/X = 0.4$, and this value decreases with increasing $r$. Such considerations are important in determining the coronal temperature from the gradient of electron densities.

Individual features, such as streamers, have been studied with the aid of an assumption concerning the depth of the feature—usually that the depth and breadth are comparable. A direct attack can be made on this problem by taking balloon observations with a sufficient time delay to allow the solar rotation to produce a stereoscopic effect. Reduction of such observations involves the assumption that the corona or feature under study is constant in time. The ultimate solution would involve simultaneous observations from a near-earth and a deep-space probe. This structural problem is very important because fans or streamers may mark the locations of the corona's prime contribution to the solar wind (see Section 7.1).

*Temperature.* The temperature of the corona is now well established as $\approx 10^6$ °K, but this is a relatively recent event. Prior to 1945, many exotic ideas were put forward to explain the existence of the coronal ions, which were ultimately identified by Grotrian and Edlén as highly ionized atoms of common constituents (e.g., Fe X). In 1945, Waldmeier summarized several different lines of argument, all of which required million-degree temperatures. It was from this time that a high-temperature corona was generally accepted. Billings has noted that "it is remarkable how many phenomena that had puzzled astronomers during the preceding century were explained by this concept."

There is much evidence now on the temperature of the corona. It can be determined from the variation of electron density with $r$ and the assumption of hydrostatic equilibrium; the simplest case also assumes an isothermal corona. Such an atmosphere with spherical symmetry has a density distribution given by

$$\frac{N_e}{N_{e,0}} = \exp\left[\frac{GM_\odot m_H \mu}{R_\odot kT}\left(\frac{1}{r} - \frac{1}{r_0}\right)\right] \tag{2.16}$$

where $G$ is the constant of gravitation, $M_\odot$ the mass of the sun, $\mu$ the mean molecular weight, $m_H$ the mass of the hydrogen atom, $R_\odot$ the solar radius, $k$ the Boltzmann constant, $T$ the temperature, $r$ the heliocentric distance in solar radii, and the subscript zero denotes a reference level. Equation (2.16) implies a linear variation between log $N_e$ and $(1/r)$ with the slope determined by the temperature and the mean molecular weight. If we take logarithms of equation (2.16), differentiate, and solve for the temperature, we find

$$T = \frac{1.00 \times 10^7 \mu}{d \log_{10} N_e/d(1/r)} \tag{2.17}$$

For a composition of 10 hydrogen atoms to 1 helium atom, $\mu = 0.608$. Plots of log $N_e$ versus $(1/r)$ for the equatorial region show a straight line behavior out to about $3R_\odot$ with a slope $d \log_{10} N_e/d(1/r) = 4.0$. This simple approach determines a temperature of $1.5 \times 10^6$ °K. Slightly lower temperatures are sometimes quoted for the polar corona, but they should be treated with caution because it is difficult to obtain data which unambiguously refer to the polar regions.

Equation (2.17) ignores any possible expansion of the corona (as implied by the solar wind) and the effects of fine structure. The equation of motion for a spherically symmetric expanding corona is

$$\frac{d\tau}{d\eta} + \tau \frac{d \ln \langle N_e \rangle}{d\eta} + \frac{\mu m_H}{kT_0} \left( R_\odot g_\odot \eta^{-2} + w \frac{dw}{d\eta} \right) = 0 \tag{2.18}$$

where $\tau = T/T_0$, $T_0$ the temperature at a reference level, $\eta = r/R_\odot$, $w$ the expansion velocity, $g_\odot$ the acceleration of gravity at the solar surface, and $\langle N_e \rangle$ the average electron density as determined from eclipse studies. The observed electron densities can be fitted with a sequence of inverse powers of $\eta$ and equation (2.18) solved for the temperature; the solution requires the specification of $T_0$ and the determination of $w \, dw/d\eta$. The solar wind flux and the constant in the equation of continuity, $N_e wr^2 = $ const., are assigned by quantities observed directly near the earth. The known densities and the equation of continuity determine the term $w \, dw/d\eta$. The integration constant $T_0$ is determined by the form of the equation and the unstable nature of the solutions; $T_0$ is very well determined simply by insisting that the temperature be neither negative nor infinite in the range 1.1 to $7R_\odot$. Sample results are shown in Figure 2.11.

The effects of the fine structure can be included by noting that part of the observed or apparent decrease in the coronal electron density is actually due to the steady decrease in the fraction of space occupied by matter, as shown in equation (2.15). Only the physical decrease in the "model rays" is relevant to the temperature determination, and the temperature is calculated by replacing $\langle N_e \rangle$ in equation (2.18) with $X\langle N_e \rangle$. These temperatures are also shown in Figure 2.11.

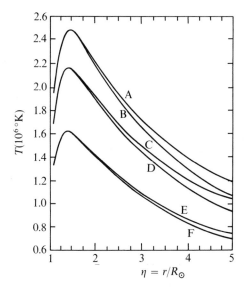

y

FIGURE 2.11

The temperature distribution as computed from electron densities under various assumptions including a flux ($N_e w$) at earth of $1 \times 10^8$ particles/cm²-sec where applicable.
(A) Inhomogeneous expanding model, H:He $=5$:1.
(B) Inhomogeneous static model, H:He $=5$:1.
(C) Inhomogeneous expanding model, H:He $=10$:1.
(D) Inhomogeneous static model, H:He $=10$:1.
(E) Homogeneous expanding model, H:He $=10$:1.
(F) Homogeneous static model, H:He $=10$:1.
(After J. C. Brandt, R. W. Michie, and J. P. Cassinelli.)

The calculated temperatures for $r < 4R_\odot$ are not greatly changed by the expansion; they are, however, significantly changed by the inclusion of the fine structure. A coronal temperature of about $2 \times 10^6$ °K would follow from these curves. The values are entirely consistent with the other methods of determining $T$ (as discussed below) except for the region just above the transition zone: the spectroscopic evidence for that region appears to require a much steeper gradient.

Spectroscopically, the temperature can be determined from measurements of the profiles of the coronal emission lines. If the line is broadened only by the Doppler effect and the corona is optically thin, the profile follows the law

$$\frac{I}{I_0} = \exp - \frac{(\lambda - \lambda_0)^2}{(\delta\lambda_0)^2} \tag{2.19}$$

where $I$ is the intensity, $\lambda$ is the wavelength, and the subscript zero refers to the line center. The Doppler width is given by

$$\delta\lambda_0 = \frac{\lambda}{c} \left( \frac{2kT}{\mu m_H} \right)^{1/2} \tag{2.20}$$

The raw observations must be corrected for the effects of the instrumental profile to determine the full width at half intensity, $h = 1.67\,\delta\lambda_0$. The temperature is then

$$T = 1.95 \times 10^{12}\,\frac{\mu h^2}{\lambda^2} \tag{2.21}$$

Observational results often quoted for the red line of Fe X ($\mu = 55.85$) at 6,375 Å are $h = 0.89$. Equation (2.21) gives a temperature of $2.1 \times 10^6$ °K. A variety of temperatures from 1.2 to $4.5 \times 10^6$ °K are found for different coronal regions; the majority of determinations fall in the range $2.0$–$2.5 \times 10^6$ °K. Billings has estimated the effect of turbulent control motions on the derived temperature and finds that the physical temperature is about $0.25 \times 10^6$ °K lower than the value calculated from equation (2.21).

The existence of highly ionized atoms also provides a means for determining temperatures. Recall that the principal coronal emission lines in the traditional wavelength region are the green line (5, 303 Å) of Fe XIV, the yellow line (5,694 Å) of Ca XV, and the red line (6,375 Å) of Fe X; the ionization potentials of these ions are 355, 820, and 235 ev, respectively. The simple observation of species with such high ionization potentials immediately suggests a very high temperature.

The circumstances that permit the direct observation of the optical coronal emission lines may be somewhat fortuitous. Ions expected for temperatures $\sim 10^6$ °K have large energy differences between the lower states, and transitions between them would fall in the extreme ultraviolet region of the spectrum. However, the ground terms of these ions are split into levels which have energy differences lying in the optical range. These levels are metastable, and the Einstein A's referring to transitions between them are $\sim 10^2\text{sec}^{-1}$; contrast this with Einstein A's $\sim 10^7\text{sec}^{-1}$ for typical permitted transitions. This fact requires a low density medium for the lines to be visible. A high density medium would de-excite the upper level by collisions during the relatively long time spent there. Fortunately, the coronal density is low enough to satisfy this condition.

The population of ions in the various stages of ionization as a function of temperature is determined by a balance of loss-and-gain processes for a given stage of ionization. The relevant processes in the corona are collisional ionization, and radiative and/or dielectronic recombination between adjacent stages of ionization and a balance is necessary for a steady state. Denote the stage of ionization by $p$, the collisional ionization coefficient by $C$, and the recombination coefficient by $R$. The balance between adjacent stages gives

$$N_e N_p R(p \rightarrow p - 1) = N_e N_{p-1} C(p - 1 \rightarrow p). \tag{2.22}$$

If there are $N$ ions in the $(p - 1)$ and $p$-th stages of ionization, then we can introduce the degree of ionization $x$ such that there are $N(1 - x)$ in the $(p - 1)$

stage and $Nx$ in the $p$-th stage. Then,

$$\frac{x}{1 - x} = \frac{C(p - 1 \to p)}{R(p \to p - 1)} \tag{2.23}$$

Note that the degree of ionization is a function only of temperature and atomic parameters and not a function of the electron density.

If the coefficients needed in equation (2.23) are available, the distribution of a given element in the various ionization stages can be computed as a function of temperature. The ratios of ionic stages must be extracted from the observations of the optical lines and from a careful discussion of the excitation conditions. Both collisional and radiative excitation are important, and a detailed calculation is required; low in the corona ($\approx 1.1 \, R_\odot$), collisional excitation dominates, but radiative excitation must still be included for numerical accuracy.

Until rather recently, the observations and available rate coefficients yielded temperatures close to $0.8 \times 10^6$ °K. However, in 1964 Burgess re-examined the calculations of the recombination coefficient and concluded that an alternate process, dielectronic recombination, was important and would change temperatures that were inferred from studies of line intensities. The older, radiative recombination coefficients were derived on the basis of the continuum electron's energy (kinetic energy + binding energy) upon recombination being given up in a photon of the same total energy. If, however, a continuum electron has an energy equal to the excitation energy of two bound states, it can excite a bound electron to one of these states and occupy the other. This yields a recombined atom with two excited electrons which cascade to lower states and emit radiation. Thus, two electrons are involved in this process, and, hence, the name, dielectronic recombination. In very hot plasmas, such as the corona, dielectronic recombination is more efficient than ordinary radiative recombination, and the recombination rate is increased by about one order of magnitude. To produce the same observed degree of ionization, the ionization rate (which increases with increasing temperature) must also increase. The analysis indicates a temperature of about $2 \times 10^6$ °K.

The ionic temperatures can be estimated directly with the aid of mass spectrometer determinations of the composition of the solar wind obtained by the Vela group (Section 5.4). Several different ionization stages of oxygen were observed, and it was easily established that the ionization balance was fixed in the corona at $r \approx 1.5 \, R_\odot$. The temperature was found to be $1.7 \times 10^6$ °K, but this value may be representative of relatively cool solar wind plasma at the earth.

Sometimes ratios of lines from different elements or widely separated stages of ionization are used to infer a temperature or a change in temperature with heliocentric distance. The ratio of the green line (Fe XIV) to red line (Fe X) intensities is often used. Unfortunately, the ionization potentials of the two

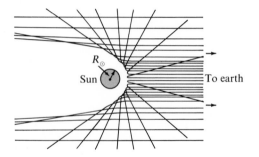

FIGURE 2.12
Ray paths in the corona at 18 Mc/s. (After
R. N. Bracewell and G. W. Preston.)

ionic species differ by 120 ev. Thus, possible coronal fine structure suggests
caution; from observation, the fine structure seems clearly to be present
because the widths of the red line consistently correspond to lower temperatures than do the widths of the green line. Apparently there are different
temperature regions on a scale too fine to be resolved by present observational
techniques. Information obtained from line ratios (such as the green line:red
line) must be regarded as primarily qualitative.

Another indication of the coronal temperature is the widths of the Fraunhofer lines in the scattered K corona. The Fraunhofer lines are generally
undetectable, with the possible exception of the H and K lines of Ca II.
Equation (2.21) can be solved for the full width at half intensity $h$ for $T =
2 \times 10^6$ °K and for the atomic weight of the electron, $\mu = 1/1,836$. These
numbers give $h \approx 170$ Å. The observations are very difficult, and present
results are consistent with any temperature greater than about $10^5$ °K.

The final " prime " method for determining the coronal temperature is the
solar radio brightness temperature for frequencies of about 18 Mc/sec.
Assuming a homogeneous corona and using Snell's law, one can calculate, the
ray paths through the corona for a medium with a varying index of refraction.
The index of refraction varies according to

$$n = \left[ 1 - \frac{e^2 N_e}{\pi m_e} \frac{1}{f^2} \right]^{1/2} \tag{2.24}$$

where $e$ is the electronic charge, $N_e$ is the electron density, $m_e$ is the mass of
the electron, and $f$ is the frequency of the radio wave. The plasma frequency is
defined by $f_0 = (e^2 N_e/\pi m_e)^{1/2}$.

The behavior of the ray paths in the corona is especially interesting. The
ray which approaches closest to the sun is the one toward the center of the
disk; this ray approaches to a distance such that $n = 0$ or $f = f_0$. Other rays
from a given observing location do not penetrate farther into the solar
atmosphere (see Figure 2.12). If one chooses to keep the wavelength close to

ten meters, the ray path can be confined entirely to the corona. If the assumption of a homogeneous corona is essentially correct, then radio observations of the corona can be carried out in the radio wavelength range without the contaminating " glare " of the photosphere. Thus, radio observations are a potentially valuable source of observations of the corona against the solar disk. Some complications are expected from the existence of the solar magnetic field, but these should be important only over large sunspots where a substantial magnetic field could be present.

The equation of transfer for radio wavelengths is derived with consideration of the change of intensity caused by the varying index of refraction. Thus, $(I/n^2)$ is constant if there are no absorbing or emitting processes. The standard equation of transfer can be integrated to yield the emergent intensity, or

$$I = \int_0^{\tau_0} e^{-\tau} B(T)\, d\tau \tag{2.25}$$

The integration is carried out along the ray path which is computed as described above. The opacity is also given by a line integral

$$\tau = \int_s^{\infty} K\, ds \tag{2.26}$$

where K is the absorption coefficient.

The emission per unit frequency in the corona is given by the Planck function,

$$B_\nu(T) = \frac{2h\nu^3}{c^2} \frac{1}{e^{h\nu/kT} - I} \tag{2.27}$$

The Planck function is permissible in the corona because only a maxwellian velocity distribution is required, not an approach to local thermodynamic equilibrium. The physical interpretation of equation (2.25) gives the emergent intensity as the emission from each point along the trajectory reduced by the intervening opacity.

In the radio wavelength range, the Planck function can be replaced by the Rayleigh-Jeans approximation, $B(T) = 2kT/\lambda^2$. Similarly, the intensity in the radio region is related to the brightness temperature $T_b$ through $I = 2kT_b/\lambda^2$. Hence, equation (2.25) can be written as

$$T_b = \int_0^{\tau_0} T e^{-\tau}\, d\tau \tag{2.28}$$

If the temperature $T_c$ is constant in the corona, then

$$T_b = T_c(1 - e^{-\tau_0}) \tag{2.29}$$

The observations would appear to determine the coronal temperature, and values close to $1 \times 10^6$ °K are cited in the literature. However, the situation is complex, especially for the optically thin case. The optical depth for thermal emission (free-free transitions) is proportional to $N_e^2/T^{3/2}$, and, hence, the emission is proportional to $N_e^2/T^{1/2}$. Thus, the problem of the coronal radio temperature is closely coupled with the problem of determining the opacity along the ray path. Most radio workers quote coronal temperatures of about $1 \times 10^6$ °K (for an optically thick corona), although considerable uncertainty exists. If such temperatures are correct, a bona fide discrepancy exists between them and coronal temperatures that have been inferred by other means. This possible discrepancy has led to a questioning of the basic equation of transfer. The problem has been hotly discussed, but the present consensus is in support of the traditional equation of transfer and holds that it is not the source of the discrepancy. Perhaps, the cause is inherent in the analysis or the result of coronal irregularities. Very recent results with greatly improved resolution give higher temperatures in essential agreement with other methods.

Finally, the temperatures can be inferred from the decay times of Type III radio bursts on meter wavelengths. The Type III bursts appear to be plasma oscillations at the plasma frequency excited by a fast group of particles (electrons) traveling through the solar atmosphere at speeds of about $c/3$ to $c/2$. The plasma frequency can be computed as a function of height in the corona from an assumed model, and thus the frequency drift with time gives the velocity of the exciting disturbance. Plasma oscillations will damp out due to collisions between ions and electrons, as $e^{-vt}$ where $v$ is the collision frequency in the plasma; the collision frequency is proportional to $N_e/T^{3/2}$. The density can be obtained because the oscillation is at the local plasma frequency [see equation (2.24)]. Hence, measurement of $v$ determines the temperature. Values of $T$ near $2.5 \times 10^6$ °K are generally quoted although there are uncertainties and exceptions. The value that was quoted is entirely compatible with the temperature expected for the corona over an active solar region.

In summary, most available evidence indicates a coronal temperature near $2 \times 10^6$ °K. No serious conflict exists at the present time.

*Composition.*   The composition of the corona influences many other coronal investigations and is an interesting topic in itself. The temperature inferred from density gradients [equations (2.17) and (2.18)] is directly proportional to the mean molecular weight in the corona, which is determined entirely by the relative abundance of helium in the corona. Exactly the same observed gradient gives a 22 percent higher $T$ for H:He = 10:1 ($\mu = 0.608$) than for a pure hydrogen corona ($\mu = 0.5$). Commonly accepted values of H:He are near 10:1, and the temperature thus derived is compatible with values obtained by other means. However, the coronal helium abundance cannot be determined directly because helium is completely ionized in the corona.

An indirect determination can be attempted by fitting the relative abundance of He in solar cosmic rays onto the photospheric (spectroscopic) abundances of the other elements. A detailed analysis gives the ratios of C, N, and O to H for the photosphere; a specific result is log $[N(O)/N(H)] = -3.23$. Cosmic ray results for He, C, N, and O are available. The charge to mass ratios of these nuclei are the same, and their relative abundance should remain constant through the flare process. Their relative abundances are reasonably constant from event to event, and the relative abundances of the C, N, and O nuclei are in reasonable agreement with the spectroscopic results. The cosmic ray ratio is $N(He)/N(O) = 107$, and the combination gives $N(He)/N(H) = 0.06$. This number is some sort of average between the photosphere and the flare location. The coronal density of helium could be higher (see Section 5.4).

An abundance of heavier elements, such as iron, can be determined from analyses of the optical lines from the ion. The absolute line intensity gives the total number of ions in the appropriate upper level along the line of sight, and a detailed calculation is involved in determining the total iron abundance. The uncertainties in such a calculation cannot be minimized, but for more than two decades many calculations have indicated an overabundance of the heavy elements relative to the photospheric abundances. Typical values show that iron is some ten to twenty times more abundant (relative to hydrogen in the corona as compared with the photosphere). This result could involve a process of diffusion or a mass dependent solar wind. This subject is discussed further in Section 5.4.

*The Magnetic Field.*   A general magnetic field in the corona can be inferred from the results of a variety of indirect methods. Recall the previous discussion of coronal fine structure and the polar plumes. The field is apparently required for support of prominences, channeling of moving prominence material, and the form of coronal material around solar-active regions. The magnetic field is assumed in many calculations concerning the theory of the corona, such as the problem of mechanical heating. The indirect evidence and the attempts to measure the coronal magnetic field have been reviewed by Billings.

Despite the prima facie nature of considerable traditional evidence, the best evidence comes from space probe measurements of $B$ near the earth. This near-earth field is apparently the photospheric field which is convected by the solar wind. This fact is established by evidence of a close correlation between the solar and interplanetary field with a time delay of about five days (Section 5.6). The quiet-time interplanetary measurement gives a radial field $B_r = 3.5\gamma(1\gamma = 10^{-5}$ gauss$)$. An inverse square variation of this component with distance implies a surface field of 1.5 gauss, a value in excellent agreement with the surface measurements. Thus, an inverse square law with a surface field of 1.5 gauss gives a good *typical* value for the field in the corona. Much higher fields can be expected in the corona over centers of solar activity. (see Section 7.1 for discussion of the possible configurations of the field in the corona.)

## 2.5    Solar Activity

Solar activity is a complex phenomenon with many possible effects on the solar wind. Here we give a descriptive summary with emphasis on broad concepts.

*Basic Manifestations.*    Sunspots have been observed since 1611. They appear as a dark central region, called the umbra, and a less dark surrounding region called the penumbra. Their appearance and structure is shown in Figure 2.13. The umbra is somewhat featureless, although some granulation-like phenomena have been observed; the umbra is composed of filaments oriented radially with respect to the center of the umbra. Diameters of sunspots measure from several thousand to several tens of thousands of kilometers. Sunspot groups, such as shown in Figure 2.14, can attain lengths of $10^5$ km or over; sunspots almost always occur in groups.

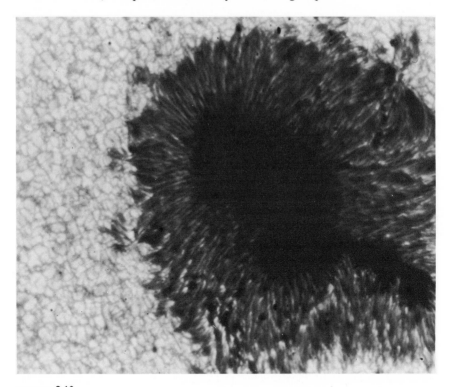

FIGURE 2.13

A balloon photograph of sunspots clearly showing the principal features. (Courtesy Project Stratoscope of Princeton University, sponsored by the Office of Naval Research, the National Science Foundation, and the National Aeronautics and Space Administration.)

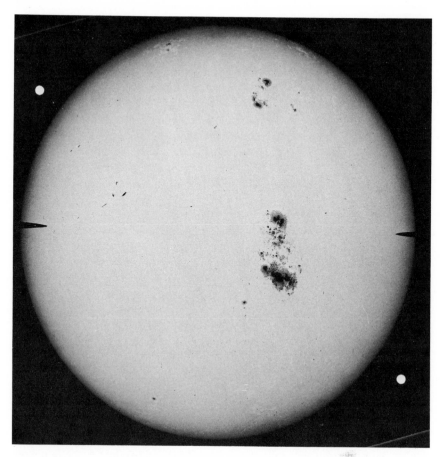

FIGURE 2.14
Large sunspot groups. (Courtesy Mount Wilson and Palomar Observatories.)

The number of spots and groups on the solar surface varies with time. A measure of sunspot activity is the Wolf relative sunspot number,

$$R = k(10g + f) \qquad (2.30)$$

where $k$ is a normalization constant assigned to an observer and his equipment, $g$ is the number of groups, and $f$ is the number of individual spots. There is obviously some arbitrariness in this definition, but the usefulness of this quantity has been unquestionable; reliable values of $R$ since about 1700 are available. Twenty-seven-day means are commonly used to remove the effects of solar rotation.

A fairly regular variation in $R$ during the eleven-year solar cycle is well known. Periods when $R > 100$ are called solar maximum; periods when

$R < 10$ are called solar minimum. Many phenomena associated with solar activity vary synchronously with the sunspot number.

Physically, sunspots are known (via the Zeeman effect) to have magnetic fields of hundreds and occasionally thousands of gauss. These fields are sufficient to seriously inhibit the convective energy transport in the regions below the photosphere. This choking effect reduces the energy supplied to the base of the photosphere and results in their cooler temperature and darker appearance. The integrated intensity for the umbral center is in the ratio $I(\text{spot})/I(\text{photosphere}) = 0.4$. Thus, the effective temperature is approximately $(5{,}750 \,^\circ\text{K}) \times (0.4)^{1/4} = 4{,}600 \,^\circ\text{K}$. This temperature is consistent with the "spectral type" of sunspots of about K3 V and the appearance of molecular bands.

The problem of the disposition of the energy "choked off" is unresolved. Traditionally, bright rings sometimes seen around sunspots have been described as manifestations of the deflected energy. However, the amount of energy in the bright rings is not equal to the amount lost; moreover, bright rings themselves are rare. The excess energy could be transported through the sunspot in the form of magnetic waves; these waves could be deposited in the corona above active regions to produce the coronal enhancements

The location of sunspots (and all phenomena associated with sunspots) on the solar disk varies with time. The migration of the sunspot zone is clearly shown in the butterfly diagram in Figure 2.15. The initial spots of a given sunspot cycle appear at solar latitudes of about $\pm 30^\circ$, reach $\pm 15^\circ$ by sunspot maximum, and appear within $10^\circ$ of the solar equator by sunspot minimum. Note that sunspots show little proper motion during their lifetime, and, hence, the migration refers only to the region of appearance.

Sunspots appear in groups of varying complexity, but detailed magnetic studies show that sunspots are only one manifestation of a larger phenomenon, the magnetic region. The sunspot simply corresponds to a region where the magnetic field is large enough to suppress convection. The evolution of the solar fields may be closely related to the supergranulation motion, as suggested by Leighton. According to him, the supergranulation itself consists of horizontal motions of about 0.4 km/sec in cells with characteristic dimension of some 30,000 km and lifetime of some $10^4$ to $10^5$ sec. The cellular motion convects the magnetic field to the cell boundaries. Since Ca II K line emission is known to correspond to regions of enhanced magnetic field, this behavior could account for the general appearance and scale of the chromospheric network (see Figure 2.16).

The supergranulation motions buffet the solar field lines and move the magnetic field lines about in a random-walk or diffusion process. A sequence of events could begin when a magnetic flux loop protrudes into the solar atmosphere and sunspots can be formed in the areas with strong field. As the buffeting proceeds, the bipolar magnetic region (BMR) formed by the emer-

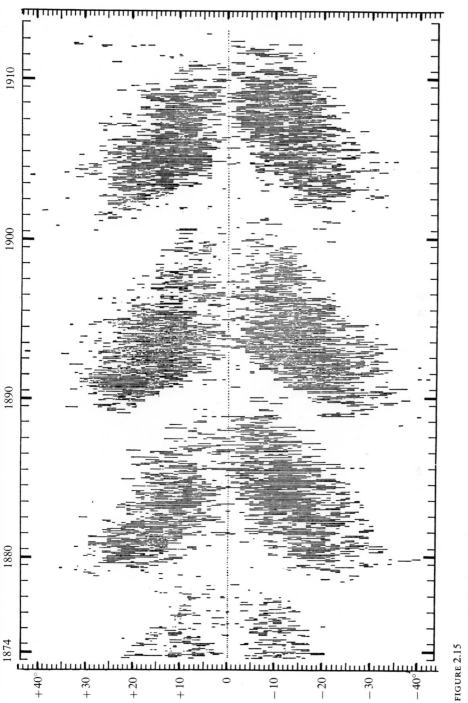

FIGURE 2.15

Maunder's butterfly diagram showing the distribution of sunspot centers in heliographic latitude as the sunspot cycle progresses. (Courtesy of the *Monthly Notices of the Royal Astronomical Society.*)

FIGURE 2.16

A spectroheliogram taken in the central emission core ($K_3$) of the Ca II K line showing plages and chromospheric fine structure. (Courtesy Observatoire de Paris-Meudon.)

gence of the flux loop increases in area at a rate of $10^4$ km²/sec. Since the net magnetic flux remains constant, the field strength decreases, the convective motions cannot be suppressed, and the spots vanish. Leighton showed that many features of the solar cycle can be explained by following the field lines and computing a field pattern that develops out of the algebraic sum of the fields associated with past sunspot groups. This kind of algebraic book-keeping accounts for the reversal of the polar field in a natural way. The general picture described here may account for the behavior of the magnetic field in the solar atmosphere. However, the question of the ultimate origin of the " atmospheric " field vis-à-vis the general solar field has yet to be answered. Solar field regions of sufficient size appear to form the " sectors " of the solar wind (see Section 5.6).

The BMRs obey the laws of polarity for sunspots discovered by Hale: (1) the preceding ($p$) and following ($f$) portions (in the sense of solar rotation) of the BMR are of opposite polarity, (2) The $p$ and $f$ portions are of opposite polarity in the northern and southern hemispheres, and (3) The $p$ and $f$ parts of the BMR in each hemisphere change polarity with each new solar cycle. Observations appear to indicate that the net magnetic flux in the $p$ and $f$ portions of a BMR are approximately equal.

The plages or faculae appear as bright regions in K line or H-$\alpha$ spectroheliograms. They are occasionally seen in white light near the limb. Such emission is closely associated with the presence of magnetic fields greater than about twenty gauss, and the plages could result from an enhancement of the process which normally heats the chromosphere. Plages are closely associated with sunspots but outlast them by one or more solar rotations.

The close association of plages and the magnetic field is emphasized by their structure and evolution. They appear to exhibit the same basic fine structure as the quiet chromosphere and could simply be areas with a high density of bright granules, which are sometimes called plagettes. (Magnetic fields of $\sim 10^2$ gauss have been observed in the plagettes.) Plages never disappear suddenly; they merge continuously into the chromospheric network. This generally diffusive process is strongly reminiscent of the process by which the supergranulation motions disperse the field lines in a magnetic region.

Prominences are seen on the limb as structures protruding into the corona and on the disk as dark "filaments"; these appearances are shown in Figure 2.17. Although the overall structure of prominences is generally constant, there is substantial motion (mostly downward) associated with the fine structure. Physically, the density of prominences is about a factor of $10^2$ higher than the surrounding corona, and the temperature is close to $10^4$ °K, or about a factor of $10^2$ lower than the corona. Hence, the corona and a prominence can be approximately in lateral pressure equilibrium. However, support for the prominence is still needed and the magnetic field is often invoked.

Prominences exist in basically two varieties: (1) active or sunspot prominences are closely associated with sunspots and occur early in the development of a sunspot group. Typical forms of active prominences are loops rooted in the spot group or condensations at about the same height in the corona. Typically, the sunspot prominences have heights $\sim 10^5$ km; and (2) quiescent prominences occur much later in the development of an active region and are usually not near sunspot groups; they are closely associated with plages. This is the kind of prominence shown in Figure 2.17. Quiescent prominences develop a blade-like appearance and tend toward an orientation parallel to the solar equator because of the solar differential rotation. Typical dimensions of a developed quiescent prominence are a length about 200,000 km, a height of about 50,000 km, and a thickness of 8,000 km.

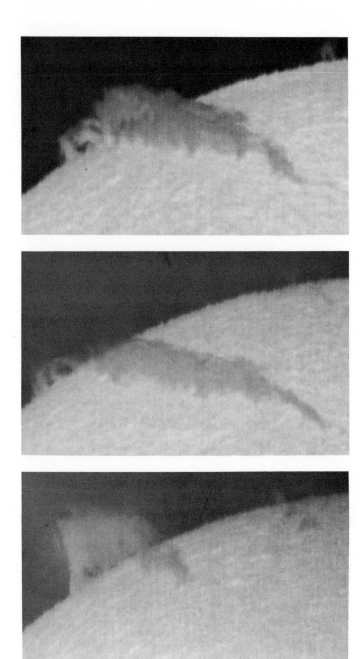

FIGURE 2.17
A photographic sequence taken on April 9, 10, and 11, 1959, top
to bottom respectively, showing the correspondence between
filaments and prominences. (Courtesy Observatoire de Paris-Meudon.)

Sunspot prominences last only a few days, and, therefore, theoretical questions of support and possible dissipation by the hot corona are not very pressing. The forms are certainly determined by the magnetic field, and a rough equilibrium could be established through a balance of the rates of condensation and dissipation or mass loss along the field lines into the chromosphere. However, quiescent prominences last for several solar rotations, and the theoretical questions require some consideration.

The question of dissipation is largely one of energy balance. Energy input to the prominences comes from the hot coronal particles that are absorbed by the prominence; this source would be reduced by the presence of a magnetic field, which would reduce the effective mean free path of the coronal particles. Energy loss occurs primarily in the Lyman continuum ($\lambda < 912$ Å), and a balance occurs for prominences with a thickness of about $10^4$ km. The existence of the magnetic field does not appear crucial to the energy balances.

However, support of quiescent prominences during the months of their existence definitely requires the magnetic field, and these prominences are found where the field is parallel to the surface. This condition can occur in a variety of locations, and it can be easily visualized when a prominence (filament) occupies the surface which separates the positive and negative parts of a BMR. Although there are some complications, this view appears to be basically correct.

If the quiescent prominences are supported by a magnetic field of sub-photospheric origin, the violent changes sometimes associated with prominences can be understood in terms of changes in the field or the currents producing it. These changes can be spectacular, and an example was shown in Figure 1.4.

Closely associated with quiescent prominences are the coronal fans discussed in Section 2.4. Other coronal regions that are associated with centers of activity are the coronal enhancements (which appear as the coronal extensions of plages) and the coronal condensations (which are associated with active sunspot regions, sunspot prominences and flares) (see Section 2.4 discussion of coronal regions).

The solar flare is the final major manifestation of solar activity. Flares are of interest not only as a problem in solar physics but also as a problem in interplanetary and planetary physics because of the effects of the release of matter and radiation. Traditionally, flares are observed in the light of H-$\alpha$ (and occasionally white light) as localized brightenings near sunspots. Flares are classified on the basis of brightness and area. The older, traditional scheme assigns an importance in increasing order of microflares, subflares ($1^-$), 1, 2, 3, and $3^+$; the lifetimes of flares vary with importance from twenty minutes for importance 1 to three hours for importance $3^+$. The new, dual importance scheme adopted by Commission 10 of the International Astronomical Union in January 1966 classifies flares by area and brightness. The area classification ranges from $S$ for flares with "corrected" or "real area" in heliographic square degrees of 2.0 or less, through 1, 2, and 3 to 4 for flares of 24.7 square heliographic degrees or greater. The intensity scale is assigned on the basis of whether the flare is faint ($f$), normal ($n$), or rather bright or brilliant ($b$). For example, a flare of exceptional brightness and a corrected area of 20.0 square degrees would be classified $3b$.

The visual spectrum of flares is similar in appearance to the flash spectrum of the chromosphere and characteristic of a gas at $\sim 10^4$ °K, but dramatic changes in the solar spectrum in the x-ray region have been recorded at the time of solar flares. Spectral scans in the wavelength region near 2 Å indicate the presence of highly ionized iron (Fe XXIII–Fe XXV) This evidence, the total amount of x-rays that were produced, and the production of solar cosmic rays imply the production of a high energy plasma; equivalent temperatures of about $10^8$ °K have been quoted in the literature. The difference in the visual and x-ray spectra of flares serves to illustrate the conceptual difference between the flare and the flare event. The flare is defined as something observed in a particular wavelength; many wavelengths contain little evidence of the high energy or high temperature aspect of the flare event. Thus, a confusion of cause and effect is possible. An attractive working hypothesis attributes the high energy particles to a flare event and the visual flare radiation to an interaction of the particles produced in the event with the ambient chromosphere.

Some ideas concerning the origin of flares can be obtained from consideration of the total energetics. Flares produce mass motions and significant radiation from the visual to the x-ray wavelength region. Estimates of the total are rather uncertain, but a value of $10^{28}$ ergs/sec for $10^3$ sec or $10^{31}$ ergs is reasonable. If this energy comes from a typical flare area of $10^{19}$ cm$^2$ and a thickness of $10^8$ cm, an energy requirement of $10^4$ ergs/cm$^3$ results. Such energy densities cannot be found in thermal motions or the bulk motions usually considered, and the traditional candidate is the magnetic field which has energy density ($B^2/8\pi$). The energy requirement can be met with a field of 500 gauss. Several models have been proposed for annihilation or conversion of the magnetic field into particle energies. On the other hand, Hyder has obtained evidence showing that infalling prominence material precedes some flares and that the energy requirement can be satisfied (at least in some cases). In view of the difficulties with the magnetic flare theories, this alternative hypothesis should be considered.

Flares are associated with considerable mass motions such as flare surges, and the regions near flares are influenced by them. Portions of the chromospheric network are brightened, but there seem to be preferred and excluded directions for this effect. Flares show considerable filamentary and other fine structure as shown in Figure 2.18.

# October 24, 1968

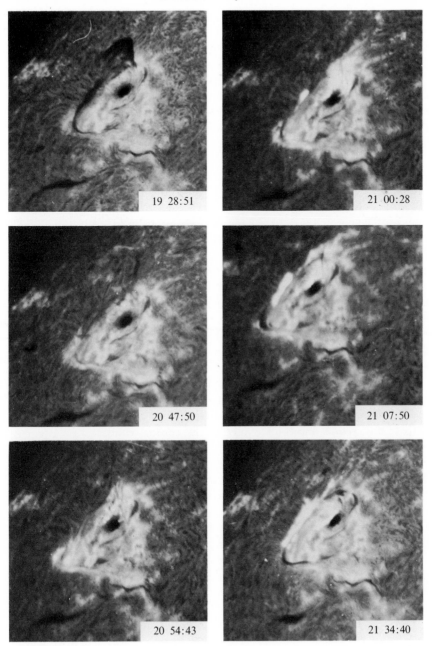

FIGURE 2.18

An H-α photographic series showing the fine structure in a solar flare. (Courtesy
H. Zirin and Mount Wilson and Palomar Observatories.)

*The Center of Activity.* The tremendous complexity of the multifarious aspects of solar activity is obvious from the preceding discussion. This fact has led to the concept of a "center of activity" (CA)—an attempt to schematize the details of solar activity in order to further understanding and to assist in concentrating the discussion on the most important points. A description of a fully developed CA is given below; note explicitly that the development of individual CAs can differ appreciably from the "typical" CA. This description is based almost entirely on visual wavelength observations. As evidence accumulates for other wavelengths such as x-rays, it must be added to the "typical" description.

Day 1: The chromospheric fine structure shows a systematic alignment around an area which develops into a plagette. Small spots or pores appear a few hours after the plagette. A magnetic region of a single polarity often appears. The coronal (green) enhancement forms after the spots appear, and the enhancement parallels the plage in development and extent.

Day 2: The plage area increases in size and brightness. The magnetic region assumes a bipolar character. The west end of the plage shows the first *p* spot. The magnetic region achieves a size of about 50,000 km. The intensity of the green line increases in the enhancement. The first signs of flare activity appear as well as the first small, unstable filaments.

Day 5: The *f* spot appears at the east end of the plage region; the two main spots slowly separate and smaller spots are found between. The active or sunspot prominences appear, and flares occur between the *f* and *p* spots.

Day 11: The plage continues to increase in size and brightness and reaches dimensions $\sim 10^5$ km. The spots reach maximum development, and the field in the magnetic region becomes variable and irregular. Flares reach peak activity and yellow-line brightenings of the coronal condensation occur in association with the flares.

Day 27: Plage continues to increase in size. The *f* spot disappears as it is bridged by normal photosphere, and usually only the *p* spot remains. The flux in the magnetic region reaches maximum as does the green line brightness in the coronal enhancement. Flares are now rare. A quiescent prominence forms on the poleward side of the plage. It points toward the *p* spot and makes an angle of about 40° with the meridian.

Day 54: There are no spots, and the brightness of the coronal green line decreases. The brightness of the plage decreases, and the area is cut in half by the prominence (filament), which is now some $10^5$ km long and swinging toward an east-west orientation.

Day 81: The length of the prominence increases, and the turning toward an east-west orientation continues. The field strength in the magnetic region declines. The plage dissolves into the bright chromospheric network.

Day 108: The plage is completely dissolved. The prominence achieves its maximum length and lies nearly parallel to the equator.

Day 135: The *CA* is undetectable in the photosphere or chromosphere except for some chromospheric fine structure oriented around the prominence (filament). The BMR occupies about one-twentieth of the disk. The length of the prominence decreases as it starts its poleward migration.

It is difficult to follow an individual CA further, but the following description is often given. It is included here as a guide to the subsequent evolution of a CA.

Day 162–270: The BMR occupies one-fifth of the solar disk, and it assumes a unipolar (UMR) character. The prominence filament joins the polar crown of filaments.

Most phenomena associated with solar activity are closely related to sunspots and hence closely follow the number and latitude variation of sunspots through the sunspot cycle.

The phenomena of solar activity are all seen to be closely associated with the presence of the solar magnetic field—hence the hypothesis that essentially all aspects of solar activity are due to an emergence of a magnetic field from the subphotospheric layers. Babcock and Leighton have done much to support this hypothesis in terms of the discussion of the evolution of such a field. Many facts related to solar activity can be explained including the existence of the polar solar field. In addition, the solar wind carries the emergent field into interplanetary space to form the observed "sectors" (Section 5.6). Results relating to the sector structure (Section 5.6) have been discussed by Wilcox and Howard in terms of an underlying pattern to the photospheric magnetic field that is not disrupted by the solar differential rotation.

# Bibliographical Notes: Chapter 2

*General reference*

References for this chapter are confined as much as possible to review articles and texts. The following references cover most subject areas of solar physics:

1. Kuiper, G. P., Ed.: *The Sun*, University of Chicago Press, Chicago (1953).
2. Ambartsumyan, V. A., Ed.: *Theoretical Astrophysics*, Pergamon Press, New York (1958).
3. Flügge, S., Ed.: *Handbuch der Physik*, **LI**, **LII**, Springer-Verlag, New York (1958–1959).
4. Aller, L. H.: *The Atmospheres of the Sun and Stars*, Ronald Press Co., New York (1963).
5. Brandt, J. C., and P. W. Hodge: *Solar System Astrophysics*, McGraw-Hill Book Co., New York (1964).
6. Zirin, H.: *The Solar Atmosphere*, Blaisdell Publishing Co., Waltham, Mass. (1966).
7. Xanthakis, J. N., Ed.: *Solar Physics*, Interscience Publishers, London (1967).
8. Brandt, J. C., and S. P. Maran: in *Introduction to Space Science*, 2nd ed., W. N. Hess and G. D. Mead, Eds., Gordon and Breach, New York (1968), p. 643.

*Section* 2.1

See the general references and the following:

9. Schwarzschild, M.: *Structure and Evolution of the Stars*, Princeton University Press, Princeton, N.J. (1968).
10. Aller, L. H., and D. B. McLaughlin, Eds.: *Stellar Structure*, University of Chicago Press, Chicago (1965).

The solar interior model follows:

11. Sears, R. L.: *Ap. J.*, **140**, 477 (1964).

The measurement of the solar oblateness is found in:

12. Dicke, R. H., and H. M. Goldenberg: *Phys. Rev. Letters*, **18**, 313 (1967).

The situation with respect to the solar neutrino flux is contained in:

13. Davis, R., D. S. Harmer, and K. C. Hoffman: *Phys. Rev. Letters*, **20**, 1205 (1968).
14. Bahcall, J. N., N. A. Bahcall and G. Shaviv: *Phys. Rev. Letters*, **20**, 1209 (1968).

*Section* 2.2

Consult the general references and the following:

15. Unsöld, A.: *Physik der Sternatmosphären*, Springer-Verlag, Berlin (1955).

16. Leighton, R. B.: *Ann. Rev. Astron. Astrophys.*, **1**, 19 (1963).

Information on velocity fields in the photosphere (granulation, etc.) is contained in:
17. Leighton, R. B., R. W. Noyes, and G. W. Simon: *Ap. J.*, **135**, 474 (1962).
18. Noyes, R. W., and R. B. Leighton: *Ap. J.*, **138**, 631 (1963).
19. Simon, G. W., and R. B. Leighton: *Ap. J.*, **140**, 1120 (1964).

A very recent paper with references to earlier work is:
20. Howard, R., A. S. Tanenbaum, and J. M. Wilcox: *Solar Phys.*, **4**, 286 (1968).

Also see:
21. Kiepenheuer, K. O., Ed.: *The Fine Structure of the Solar Atmosphere*, Franz Steiner-Verlag, Wiesbaden (1966).
22. Bray, R. J., and R. E. Loughhead: *The Solar Granulation*, Chapman and Hall Ltd., London (1967).

*Section* 2.3
See the general references. The chromosphere is reviewed by:
23. Pagel, B. E. J.: *Ann. Rev. Astron. Astrophys.*, **2**, 267 (1964).

The historical suggestions for coronal heating are in:
24. Alfvén, H.: *Monthly Notices Roy. Astron. Soc.*, **107**, 211 (1947).
25. Biermann, L.: *Zs. f. Astrophys.*, **25**, 161 (1948).
26. Schwarzschild, M.: *Ap. J.*, **107**, 1 (1948).

Recent work on coronal heating is found in:
27. Stein, R. F.: *Ap. J.*, **154**, 297 (1968).
28. D'Angelo, N.: *Ap.. J.*, **154**, 401 (1968).

A physical discussion of Landau-damping is found in:
29. Jackson, J. D.: *Classical Electrodynamics*, John Wiley, New York (1963), p. 340.

The coronal energy balance is discussed in:
30. Brandt, J. C., R. W. Michie, and J. P. Cassinelli: *Icarus*, **4**, 19 (1965).

The extent of coronal heating is discussed in Reference 30 and:
31. James, J. C.: *Ap. J.*, **146**, 356 (1966).
32. Brandt, J. C.: *Ap. J.*, **149**, 447 (1967).
33. Wentzel, D. G., and A. B. Solinger: *Ap. J.*, **148**, 877 (1967).

The variations of radiative losses are calculated in:
34. Weymann, R.: *Ap. J.*, **132**, 380 (1960).

*Section* 2.4
Theory and observations of the corona are summarized in the general references and in:
35. Evans, J. W., Ed.: *The Solar Corona*, Academic Press, New York (1963).
36. Shklovskii, I. S.: *Physics of the Solar Corona*, Addison-Wesley Publishing Co., Reading, Mass. (1965).

37. Billings, D. E.: *A Guide to the Solar Corona*, Academic Press, New York (1966).
38. Newkirk, G.: *Ann. Rev. Astron. Astrophys.*, **5**, 213 (1967).

Influences of coronal expansion and inhomogeneity on the coronal temperature inferred from density gradients are calculated in Reference 30 and in:
39. Brandt, J. C., R. W. Michie, and J. P. Cassinelli: *Ap. J.*, **141**, 809 (1965).

The importance of dielectronic recombination in the corona is described in:
40. Burgess, A.: *Ap. J.*, **139**, 776 (1964).
41. Tucker, W. H., and R. J. Gould: *Ap. J.*, **144**, 244 (1966).

Solar radio astronomy is reviewed in:
42. Kundu, M. R.: *Solar Radio Astronomy*, Interscience Publishers, New York (1965).
43. Aarons, J., Ed.: *Solar System Radio Astronomy*, Plenum Press, New York (1965).

The equation of radiative transfer for a dispersive medium such as the corona is discussed in:
44. Zhelegnyakov, V. V.: *Ap. J.*, **148**, 849 (1967).
45. Pomraning, G. C.: *Ap. J.*, **153**, 321 (1968).

The solar helium abundance from cosmic ray studies is summarized in:
46. Durgaprasad, N., C. E. Fichtel, D. E. Guss, and D. V. Reames: *Ap. J.*, **154**, 307 (1968).

The problem of coronal abundances and the solar wind is discussed in:
47. Jokipii, J. R.: in *The Solar Wind*, R. J. Mackin and M. Neugebauer, Eds., Pergamon Press, New York (1966), p. 215.

48. Brandt, J. C.: *Ap. J.*, **143**, 265 (1966).
49. Delache, P.: *Ann. d'Ap.*, **30**, 827 (1967).

*Section* 2.5
Descriptions of solar activity are found in many of the general references and in:

50. Tandberg-Hanssen, E.: *Solar Activity*, Blaisdell Publishing Co., Waltham, Mass. (1967).
51. Kiepenheuer, K. O., Ed.: *Structure and Development of Solar Active Regions*, D. Reidel Publishing Co., Dordrecht-Holland (1968).

Specific aspects of solar activity are covered in:
52. Babcock, H. W.: *Ann. Rev. Astron. Astrophys.*, **1**, 41 (1963).
53. Bray, R. J., and R. E. Loughhead: *Sunspots*, Chapman and Hall Ltd., London (1964).
54. Smith, H. J., and E. v. P. Smith: *Solar Flares*, The Macmillan Co., New York (1963).
55. Severny, A. B.: *Ann. Rev. Astron. Astrophys.*, **2**, 363 (1963).

56. Hess, W. N. Ed.: *The Physics of Solar Flares,* National Aeronautics and Space Administration (NASA SP-50), Washington, D.C. (1964).

The multifarious indices and classifications used to describe solar activity are contained in the monthly publication *Solar-Geophysical Data* issued by the Environmental Sciences Services Administration (ESSA), Boulder, Colorado. A description is contained in a supplement to the February 1967 issue.

The underlying magnetic pattern in the photospheric field is discussed in:
57. Wilcox, J. M., and R. Howard: *Phys. Rev. Letters,* **20**, 1252 (1968).
58. Wilcox, J. M., and R. Howard: *Solar Phys.,* **5**, 564 (1968).

# 3

# *Basic Theory*

This chapter covers the development of the theoretical approach to the solar wind, beginning with the original hydrodynamical model of E. N. Parker. The evolution of theoretical models is traced, with some attention being paid to the inclusion of the magnetic field. Finally, the extent of the solar wind into space is discussed in terms of its interaction with the interstellar medium.

## 3.1 Parker's Hydrodynamic Theory (1958)

Parker began his paper by developing two points: (1) there was considerable evidence from studies of comet tails and from geomagnetic studies that favored a continuous expulsion of solar matter at speeds of hundreds of km/sec, and (2) Chapman's static model had a much larger pressure at infinity than could be balanced by the pressure of the interstellar medium; hence, it was postulated the interplanetary medium must expand. (This point has been discussed in Section 1.3.)

Parker proposed that the expansion was a natural result of the high temperature of the corona, and to illustrate this he worked out the first hydrodynamic model, based on a scalar pressure. The corona is assumed to be spherically symmetric, inviscid, and in a steady state; the sun is assumed to be nonrotating and to have no magnetic field. The basic hydrodynamic equations are: (1) the equation of motion

$$N\mu m_H w \frac{dw}{dr} = -\frac{d}{dr}(NkT) - \frac{GN\mu m_H M_\odot}{r^2} \qquad (3.1)$$

and (2) the equation of continuity

$$N(r)w(r)\,r^2 = N_0 w_0 a^2 = C \qquad (3.2)$$

Here,

$N$ = the total particle density,
$m_H$ = the mass of the hydrogen atom ($1.67 \times 10^{-24}$g),
$\mu$ = the mean molecular weight,
$w$ = the radial expansion velocity,
$r$ = the heliocentric distance,
$k$ = Boltzmann's constant ($1.4 \times 10^{-16}$erg/°K),
$T$ = the temperature,
$G$ = the gravitational constant ($6.67 \times 10^{-8}$ cm$^3$/g-sec$^2$),
$M_\odot$ = the mass of the sun ($1.99 \times 10^{33}$g).

The notation introduced by Parker was

$$\xi = r/a \qquad (3.3)$$

$$\tau = T(r)/T_0 \qquad (3.4)$$

$$\lambda = \frac{G\mu m_H M_\odot}{akT_0} = \frac{(V_{esc})^2}{U^2} \qquad (3.5)$$

$$\psi = \frac{\mu m_H w^2}{kT_0} = \frac{w^2}{U^2} \qquad (3.6)$$

In all of the above equations, the subscript zero refers to the quantity evaluated at the base of the corona $r = a$; and $U = (2kT_0/m_H)^{1/2}$ is the most probable velocity of protons in a maxwellian distribution at temperature $T_0$. The density can be eliminated from equation (3.1) using equation (3.2) and the resulting equation written in terms of the variables defined by equations (3.3) through (3.6). This gives

$$\frac{d\psi}{d\xi}\left(1 - \frac{\tau}{\psi}\right) = -2\xi^2 \frac{d}{d\xi}\left(\frac{\tau}{\xi^2}\right) - \frac{2\lambda}{\xi^2} \qquad (3.7)$$

Although the detailed nature of the solution requires the specification of $T(r)$, the basic form is evident from inspection of equation (3.7). Equation (3.8) shows that the right-hand side of equation (3.7) has a zero. The left-hand side can become zero either by having $d\psi/d\xi = 0$ or by having $\tau = \psi$. This feature suggests a dual nature of the solutions. Low in the corona, $d\psi/d\xi$ is positive as the plasma accelerates. Hence, if the equation is satisfied by $d\psi/d\xi = 0$, the velocity will decrease thereafter. If the equation is satisfied by $\tau = \psi$ and not $d\psi/d\xi = 0$, then $\psi$ will increase outward.

Parker set $\tau = 1$ (constant temperature) to approximate the conditions appropriate to the high thermal conductivity of the corona. If thermal conduction alone was insufficient to maintain a constant temperature, energy deposition could be invoked. Then, with the r.h.s. of equation (3.7) equal to zero or

$$2\xi^2 \frac{d}{d\xi}\left(\frac{1}{\xi^2}\right) + \frac{2\lambda}{\xi^2} = 0 \tag{3.8}$$

this reduces to

$$\frac{\lambda}{\xi^2} = \frac{2}{\xi} \tag{3.9}$$

Since the l.h.s. of equation (3.9) is larger than the r.h.s. low in the corona and also decreases faster with increasing distance, the two sides can be equal to produce a zero in the r.h.s. of equation (3.7). This occurs at

$$\xi = \frac{\lambda}{2} = \frac{G\mu m_H M_\odot}{2akT_0} \tag{3.10}$$

Parker chose the solution for which $\tau = \psi$. This latter condition implies $\mu m_H w^2 = kT$, and, hence, the thermal and bulk kinetic energies are approximately equal at the point defined by equation (3.10), which is called "the critical point." The velocity at this point is

$$w = \left(\frac{kT}{\mu m_H}\right)^{1/2} \tag{3.11}$$

Since this is close to the velocity of sound,

$$v_s = \left(\frac{\gamma P}{\rho}\right)^{1/2} = \left(\frac{5kT}{3\mu m_H}\right)^{1/2} \tag{3.12}$$

the critical point is sometimes loosely referred to as the "sonic point" and the solutions advocated by Parker as the "supersonic solutions." Choice of the solutions for which $d\psi/d\xi = 0$ at the critical point would limit the expansion

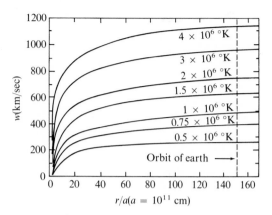

FIGURE 3.1

Parker's solar wind solutions (isothermal corona) based on Equation (3.13). (After E. N. Parker.)

velocities to less than the value given by equation (3.11) or 180 km/sec for $T = 2 \times 10^6$ °K. Such velocities could not reproduce Biermann's observations.

The solution of equation (3.7) for the isothermal case such that $\psi = 1$ at $\xi = \lambda/2$ is given by

$$\psi - \ln \psi = -3 - 4 \ln \frac{\lambda}{2} + 4 \ln \xi + \frac{2\lambda}{\xi} \qquad (3.13)$$

As one moves away from the sun, the dominant terms are

$$\psi \approx 4 \ln \xi \qquad (3.14)$$

and, hence, the isothermal region must be terminated at some finite distance. Figure 3.1 shows solutions calculated from equation (3.13). If we seek solutions which reach a velocity of 500 km/sec, this velocity is reached at $r = 5a$ for $T_0 = 3 \times 10^6$ °K, $r = 16a$ for $T_0 = 2 \times 10^6$ °K, and $r = 36a$ for $T_0 = 1.5 \times 10^6$ °K. A cutoff distance of some 20a would be quite reasonable. Note that some simplifications were introduced in the original calculations, such as setting the solar radius $= a = 10^{11}$ cm.

Thus, Parker showed that velocities $\sim 10^3$ km/sec could result from coronal temperatures existing over an extended region around the sun. The outflowing gas suggested by Biermann was then explainable in terms of the high temperature of the corona. The density at the earth's orbit was approximately 500 electrons/cm$^3$ on the early models; this value is now known to be too high by a factor of about $10^2$. The discrepancy results apparently from the assumed temperature distribution and/or filamentary structure in the corona.

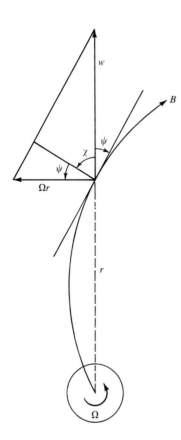

FIGURE 3.2

Schematic of the spiral form of
the interplanetary magnetic field
and the quantities used to
describe it.

Note that the supersonic solutions obtained by Parker cannot occur if the
coronal temperature is too high. This can be seen by solving for $T_0$ in equation
(3.10) for $\xi = 1$ to obtain $T_0 = 4 \times 10^6$ °K. A higher $T_0$ would displace the
critical point below the solar surface. The physical reason for this limit is
discussed in Section 3.3.

Parker also discussed the form of the magnetic field if it is carried into space
by steady expansion but with the roots of the field lines fixed on a rotating
sun. Assume that the solar wind plasma cannot cross the magnetic field lines,
i.e., that the field lines are frozen-in. The interplanetary field lines then
connect all plasma emitted from the same location on the rotating sun and
have an "Archimedes" spiral configuration as shown in Figure 3.2.

The spiral field pattern rotates with the solar (twenty-seven-day) period
while the plasma moves strictly radially. This apparent contradiction can be
visualized by imagining the grooves of a phonograph record and the needle
(as suggested by Ahluwalia and Dessler). Just as the grooves determine the
path of the needle, so does the plasma determine the shape of the field. The
condition for co-rotation on this picture follows from quantities defined in

Figure 3.2 and is $w \sin \psi = \Omega r \cos \psi$. Geometrical situations such as kinks or flare-induced changes in the field cannot co-rotate with the sun. Note that the concept of co-rotation used to describe a feature, in this case, the magnetic field, which does *not* involve azimuthal motion of material must be carefully distinguished from physical co-rotation which *does* involve azimuthal motion of material (Section 3.7). The question of the electric field and other problems associated with co-rotation of the magnetic field are discussed in Section 3.7.

If $\Omega$ is the solar rotation rate and $w$ is taken as constant, we immediately have

$$\phi - \phi_0 = \Omega t = \Omega \frac{(r_0 - r)}{w} \tag{3.15}$$

Referring to Figure 3.2 again, we see that the "Archimedes spiral angle" $\psi$ is is defined by

$$\tan \psi = - \frac{(r - r_0)\, d\phi}{dr} \tag{3.16}$$

Since $-d\phi/dr = \Omega/w$ from equation (3.15), we have

$$\tan \psi = \frac{(r - r_0)\Omega}{w} \tag{3.17}$$

The quantity $r\Omega$ is the linear velocity that corresponds to rigid body solar rotation at the earth and is approximately 430 km/sec. Thus, if $w \approx 430$ km/sec, the interplanetary magnetic field lines near the earth should make an angle of 45°, or 135° with the radius vector. Such behavior has been observed, and is discussed in Section 5.5.

In the preceding discussion we assumed that the field lines were frozen-in to the plasma. This can be calculated from the magnetic Reynolds number

$$R_M = [4\pi\sigma\, l^2]/[l/w] = 4\pi\sigma\, lw \tag{3.18}$$

where $l$ is a characteristic dimension and the conductivity $\sigma = 2 \times 10^{-14}$ $T^{3/2}$ (c.g.s. units). Physically, the magnetic Reynolds number is the ratio of the time required for a magnetic field to diffuse through a distance $l (\tau_{\text{dif}} = 4\pi\sigma\, l^2)$ to the time required for the field to be transported through a distance $l$ by bulk motions $(\tau_{\text{trans}} = l/w)$. If the Reynolds number is very large, the field is frozen-in. For $w = 5 \times 10^7$ cm/sec, $\sigma \approx 10^{-6}$ (for $T \sim 10^5$ °K), and $l = 10^{13}$ cm, we obtain $R_M \sim 10^{16}$ and the field is certainly frozen-in.

Lastly, the assumption of a scalar pressure [implicit in the form of equation (3.1)] required justification. Certainly the mean free path for protons should be less than the typical dimension and can be calculated in a proton-electron gas from the 90° deflection time of plasma physics $(t_D)$ and the mean proton speed for a gas at temperature $T$ to obtain

$$\Lambda = (1/2) \times 10^{-9} \frac{T^2}{N_e} \qquad \text{(a.u.)} \qquad (3.19)$$

Near the earth, a $T$ of $10^5$ °K and an $N_e$ of 5/cm³ are appropriate and $\Lambda = 1$ a.u. If the heliocentric distance is used as the typical dimension, $\Lambda/r = 1$. At 2.5 $R_\odot$, appropriate parameters are $T = 1.5 \times 10^6$ °K and $N_e = 10^6$/cm³ for a $\Lambda$ of $10^{-3}$ a.u. Here, $\Lambda/r = 12$. Hence, a scalar pressure is a good approximation in the inner and medium corona, but not for the distant reaches of the corona. This conclusion is apparently modified by the presence of the interplanetary magnetic field which can couple particles together and effectively provide collisions.

Objections to Parker's theory were immediate and widespread. Besides many nonrational criticisms, three substantial points against the theory were made: (1) the high fluxes found from the model implied flow velocities of hundreds or even thousands of km/sec in the corona, and effects should be seen in coronal line profiles. This point is well taken, and it is now known that the flux on Parker's model is too high by a factor of $10^2$. A less severe temperature distribution produces a gentler acceleration and a lower flux; (2) the form of the solution is determined by a unique integration constant. Changes in the coronal conditions could change this constant and render the solution unstable; and (3) the unique supersonic solution is *not* demanded by the equations of the problem. Objections (2) and (3), which are concerned with the form of the solutions, are discussed in Section 3.2.

## 3.2  The Form of Solutions: Solar Wind Versus Solar Breeze

In 1961 Chamberlain combined an energy equation, namely the first law of thermodynamics, with the equation for heat conduction; this became the third equation of the problem in which the equations of motion and of conductivity were the first and second. The new equation which resulted was

$$\frac{1}{r^2} \frac{d}{dr} \left( \kappa r^2 \frac{dT}{dr} \right) = -kTw \frac{dN}{dr} + \frac{3}{2} Nkw \frac{dT}{dr} \qquad (3.20)$$

where most of the symbols are as above and the thermal conductivity is

$$\kappa = \kappa_0 T^{5/2} \text{ erg/cm-sec-°K} \qquad (3.21)$$

Here, $\kappa_0$ is approximately $5 \times 10^{-7}$ (c.g.s.). Equation (3.20) states that the energy flow per second into a volume goes into increasing the volume or into increasing the internal energy. The time derivative is evaluated using

$$\frac{d}{dt} = \frac{\partial}{\partial t} + \mathbf{w} \cdot \mathbf{V} \tag{3.22}$$

This equation refers the net energy gain to the mass flow. Equation (3.20) contains the assumption of energy flow into a volume only by conduction; radiative losses or mechanical energy deposition are ignored. Thus, unless otherwise noted, use of equation (3.20) implies that the coronal heating takes place only in a thin shell at the base of the corona.

Chamberlain also standardized some of the notation. Besides $\tau = T/T_0$, and $\psi = w^2/U^2$, let

$$\lambda(r) = \frac{GM_\odot \mu m_H}{kT_0 r} = \frac{v_{esc}^2}{U^2} \tag{3.23}$$

Here $v_{esc}$ is the escape velocity for protons (see Section 1.2).

Equations (3.1) and (3.20) can be written in the new notation, combined, and the density eliminated using equation (3.2) to obtain

$$\frac{d}{d\lambda}\left(\frac{\psi}{2} - \lambda + \frac{5}{2}\tau\right) = -\frac{A}{2}\frac{d}{d\lambda}\left(\tau^{5/2}\frac{d\tau}{d\lambda}\right) \tag{3.24}$$

The constant A is given by

$$A = \frac{2\kappa_0 r_0 \lambda_0}{kC} = \frac{2\kappa_0 GM_\odot m_H \mu}{k^2 T_0 C} \tag{3.25}$$

Equation (3.24) integrates immediately to give

$$\frac{1}{2}\psi - \lambda + \frac{5}{2}\tau = \varepsilon_\infty - \frac{A}{2}\tau^{5/2}\frac{d\tau}{d\lambda} \tag{3.26}$$

The left-hand side of this equation contains the kinetic energy, the potential energy, the internal energy ($\frac{3}{2}\tau$), and the potential for adiabatic expansion ($\tau$). This total kinetic and potential energy per particle is equal to the energy at infinity ($\varepsilon_\infty$) less the energy still to be gained by conduction as the particle moves from a given point to an infinite distance.

The disposition of $\varepsilon_\infty$ is crucial for determining the physical nature of the flow. Chamberlain presented arguments for $\varepsilon_\infty = 0$ and numerically integrated the solar wind equation to obtain his "solar breeze" models. The parameters of this model near the earth are $w = 18$ km/sec, $T \approx 20{,}000\,°K$, and $N_e = 30/cm^3$.

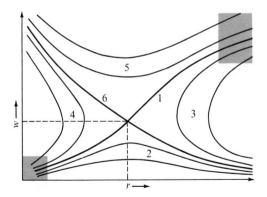

FIGURE 3.3
Different solutions to the "solar wind" equations
showing the critical point. (After E. N. Parker.)

Consider again equation (3.7); the basic nature of the solutions is not altered by the introduction of the temperature gradient. The full family of solutions is shown in Figure 3.3. The transonic solar wind solution is curve (1); this curve has $\psi = \tau$ at the critical point. The solar breeze solutions are labeled (2); these curves have $d\psi/d\xi = 0$ at the critical point. Solutions (3) and (4) are double-valued and, hence, unphysical. Solutions (5) and (6) start out super-sonic and, therefore, are also unrealistic. Empirical evidence indicates that the solution realized in practice connects the two shaded areas in Figure 3.3. Only the solar wind solutions satisfy this criterion, and it is clear that the solar breeze solutions are not realized in practice.

Nevertheless, the solar breeze curves represent perfectly well-behaved solutions to the wind equations, and they show the property $P \to 0$ as $\lambda \to 0$ $(r \to \infty)$. Consideration of the effects of the solar magnetic field on the solar wind (Section 3.7) provides a physical rather than empirical argument for choosing the solar wind or transonic curve.

Chamberlain noted that within the framework of a nonmagnetic corona heated by conduction, equation (3.26) provides an empirical criterion for solar breeze versus solar wind solutions. A solar wind solution requires positive total energy at infinity or $\varepsilon_\infty > 0$. For this to be the case, we must have everywhere $A > 2\lambda/(\tau^{5/2} \, d\tau/d\lambda)$. Thus, if conduction is sufficiently efficient, a solar wind can be attained. Physically, the limit results from the fact that the kinetic energy required for a supersonic solar wind increases with increasing density while conduction (which supplies the energy) is density independent. Thus, for a sufficiently high density, conduction cannot supply enough energy and the supersonic solution cannot be achieved. Noble and Scarf's model (discussed below) satisfied the criterion quoted, but Chamberlain's solar breeze models did not.

Chamberlain also argued that the hydrodynamic and evaporative models should be in reasonable agreement. His evaporative model had an expansion velocity and density comparable with the solar breeze solution while the temperature was about an order of magnitude higher. The circumstances that were involved in the selection of such solutions a decade ago were complex; in retrospect, however, it seems that the choice could have been adversely influenced by the approach taken to the problem. Chamberlain was thoroughly familiar with the evaporative theory, and Parker was seeking a theoretical model to explain Biermann's comet tail observations. The reader is invited to examine the papers forming part of this lively controversy. Finally, it should be noted in passing that Chamberlain's point concerning the basic similarity of the hydrodynamic and evaporative models actually was correct, but the evaporative model required modification in order to be applicable to the solar corona (see discussion in Section 3.6.).

As noted earlier, the solar wind model first presented by Parker contained an undesirable feature: very high fluxes near the earth and very high expansion velocities in the corona. In 1963 Noble and Scarf integrated the solar wind equations given by Chamberlain from the earth inward for the case of small but finite $\varepsilon$. The parameters adopted at the earth, $N_e = 3.4/\text{cm}^3$, $w = 352$ km/sec, and $T = 2.8 \times 10^5$ °K, are reasonable representations of the observed solar wind, and the computer solution also showed reasonable agreement with the coronal densities. Thermal conduction alone beyond about $2R_\odot$ is sufficient gradually to accelerate the coronal plasma through the critical point and produce a supersonic expansion at large distances.

Investigators have probed the question of the stability of the solar wind solutions, but the problem is complex and some of the earlier conclusions are now being questioned. Recently Jockers investigated the stability of an isothermal, spherically symmetric solar wind to velocity and pressure perturbations which are applied at the inner boundary. He did find stable solutions but urged further investigation.

Many basic problems are not yet understood, such as the degree to which fluid behavior is approximated in an essentially collisionless plasma. But one puzzling and surprising aspect of the inviscid hydrodynamic models described in this section (see also the references) is the fact that they are in better agreement with the observations than models with viscosity (see Section 3.4).

## 3.3   The De Laval Nozzle Analogy

The process by which the solar corona with typical thermal speeds of 180 km/sec is able to move material against the solar gravitational field (such that the escape velocity is $\approx 500$ km/sec) and give it a velocity of 400–500 km/sec and beyond may at first sight seem somewhat mysterious, particularly to

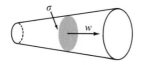

FIGURE 3.4

Quantities related to the flow
in tubes.

workers with a background in astronomy. Analagous processes have been
known to aerodynamicists and aircraft engineers for some time. Here we
explore the physics of supersonic expansion with the help of a nozzle analogy.

Consider a flow of fluid down a tube with cross section $\sigma$ (Figure 3.4). The
equation of continuity becomes

$$\sigma \rho w = \text{const} \tag{3.27}$$

The equation of motion (3.1) neglecting gravity becomes

$$- \rho w \, dw = dP \tag{3.28}$$

where $P$ is the pressure. This equation can be rewritten as

$$\frac{dP}{\rho} = \frac{dP}{d\rho} \frac{d\rho}{\rho} = -w \, dw. \tag{3.29}$$

By specifying the physical nature of the process, we can specify $dP/d\rho$. If the
flow is isothermal,

$$\frac{dP}{d\rho} = \frac{v_s^2}{\gamma} \tag{3.30}$$

where $v_s$ is the speed of sound (see equation (3.12)). If the flow is adiabatic,

$$\frac{dP}{d\rho} = v_s^2 \tag{3.31}$$

Here the latter assumption is adopted and equation (3.29) becomes

$$\frac{d\rho}{\rho} = -\frac{w}{v_s^2} \, dw \tag{3.32}$$

Taking logarithms of equation (3.27), differentiating, and evaluating $d\rho/\rho$
from equation (3.32) yields

$$\frac{d\sigma}{\sigma} = \left(\frac{w^2}{v_s^2} - 1\right) \frac{dw}{w} \tag{3.33}$$

This is a fundamental equation. Consider the various possibilities.

Case I: Here the tube is converging and hence $d\sigma/\sigma$ is negative. If the velocity is increasing, $dw/w$ is positive, and $[(w^2/v_s^2) - 1]$ must be negative. Thus $w < v_s$, and the speed of sound is the limiting flow speed in a converging tube.

Case II: Here the cross section of the tube is constant, and $d\sigma/\sigma = 0$. If the velocity is increasing, the relation requires $w = v_s$.

Case III: The tube is diverging and hence $d\sigma$ is positive. If a positive $dw/w$ is required, then $w > v_s$ is indicated.

Thus, a flow can be (1) accelerated to the velocity of sound in a converging tube, (2) accelerated through the velocity of sound in a tube of constant cross section, and (3) accelerated supersonically in a diverging tube. Such a sequence constitutes the basic principle of the de Laval nozzle or rocket engine shown schematically in Figure 3.5. The basic physical process in both the solar wind and the rocket engine is to convert random motions into directed motions. Compare equations (3.7) and (3.33); they are remarkably similar in form. In the solar wind case, solar gravity confines the hot plasma, allows the existence of the critical point, and produces the supersonic expansion. In the rocket engine, the chamber walls provide the confinement; the critical point is located in the throat of the nozzle, and the supersonic expansion results.

The role of gravity and the importance of "confinement" cannot be over-emphasized. Consider a subsonic flow in a diverging tube; equation (3.33) dictates a negative $dw/w$, and the velocity of expansion continuously decreases. Here there is no throat. Examine once again equations (3.7) to (3.9), and note that the zero in the right-hand side of equation (3.7) depends on the solar gravity; if it is too low, the zero does not exist and the solution cannot pass through the critical point. This fact is the physical origin of the result (presented in Section 3.1) that supersonic expansion cannot occur in a spherical geometry for temperatures greater than about $4 \times 10^6$ °K. For higher temperatures, $\lambda$ [equation (3.5)] is too small, the zero cannot occur, etc. This upper limit could be altered by a different geometry, formed for example by magnetic structures above prominences.

Some theoretical results from the rocket problem may be of interest. The maximum exhaust velocity from a de Laval nozzle is $(3)^{1/2}v_s$; for a $2 \times 10^6$ °K corona, the sound speed is 235 km/sec, and the maximum exhaust speed is 406 km/sec. Gravity is not included in this calculation, but the efficiency could be greatly increased by heating beyond the throat (where the gas would begin to cool adiabatically); this is the principle of the afterburner. If, on the other hand, the sound speed is not reached in the throat, the gas decelerates in the diverging section. Whether or not this occurs depends on the pressure of the background medium; the flow into a vacuum is supersonic.

In summary, the solar wind flow, as an important phenomenon with the observed properties noted above, exists only for a limited temperature range. If $T$ is too high, there is effectively no throat, and supersonic expansion

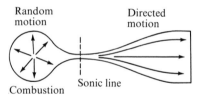

Random          Directed
motion          motion

Sonic line
Combustion

FIGURE 3.5

The de Laval nozzle or rocket engine
showing the conversion of random
kinetic motions into directed motions.

cannot occur. If $T$ is too low, the situation is essentially static—like a heavy
gas in a planetary atmosphere. For the intermediate range with $T \approx 2 \times 10^6$
°K, the supersonic flow occurs in a manner physically similar to the de Laval
nozzle in a rocket engine.

## 3.4   Solutions with Viscosity

The Navier-Stokes equation for a compressible fluid in a steady state is

$$\rho \mathbf{w} \cdot \nabla \mathbf{w} = -\nabla P - \rho \left[ \nabla \phi - \frac{w_\phi^2}{r} \mathbf{e}_r \right] \tag{3.34}$$

$$+ \eta^* \nabla^2 \mathbf{w} + (\zeta + \tfrac{1}{3} \eta^*) \nabla (\nabla \cdot \mathbf{w})$$

Here $\phi = - GM_\odot/r$; $w_\phi$ is the azimuthal velocity; $\mathbf{e}_r$ is a unit radial vector;
the viscosity, $\eta^* = 1.2 \times 10^{-16} T^{5/2}$ gm/cm-sec; $\zeta$ is the second coefficient of
viscosity. For an inviscid fluid, spherical symmetry, and $w_\phi = 0$, equation
(3.34) reduces to equation (3.1).

The centrifugal force term $\rho(w_\phi^2/r)\mathbf{e}_r$ in equation (3.34) can be roughly
estimated by means of an approximation used in studying planetary interiors
where the centrifugal force is averaged over a sphere. This averaged force
$2\rho w_\phi^2/3r$ is unimportant both for rigid body rotation in the corona and also
for realistic values of $w_\phi$ throughout the solar wind region; this term is
included, however, for a model given in Section 3.7.

Studies of the viscous terms in equation (3.34) have been undertaken by
several investigators. Normally, the second coefficient of viscosity is set equal
to zero. The term $\eta^* \nabla^2 \mathbf{w}$ is the usual viscous term while the term $(1/3) \eta^* \nabla$
$(\nabla \cdot \mathbf{w})$ arises from the fact that the gas is compressible ($\nabla \cdot \mathbf{w} = 0$ for an incompressible fluid). Several solar wind models with viscosity have been derived;

they have the common property of giving a very low expansion velocity near the earth of about 165 km/sec. Large viscous stresses are found in these models. These viscous solutions have the computational advantage of having no singularities corresponding to the critical point of the inviscid solution. This can be seen by solving equation (3.7) for $d\psi/d\xi$; the viscous terms (when added) prohibit a zero in the denominator of the resulting expression.

Meyer and Schmidt have suggested that some insight into the physical situation can be obtained through consideration of the viscous stress $T$. The sum of the viscous terms in equation (3.34) for spherical symmetry can be evaluated from the expression

$$\frac{1}{r^3}\frac{d}{dr}(r^3 T) \tag{3.35}$$

where the usual expression for the viscous stress is

$$T = \frac{4}{3}\eta^*\left[\frac{dw}{dr} - \frac{w}{r}\right] \tag{3.36}$$

The stress is composed of a part resulting from the velocity gradient and a part $(w/r)$ produced by the lateral momentum transfer in a radially diverging flow. The viscous force associated with the lateral momentum transfer has a braking effect and is important even far from the sun.

Meyer and Schmidt have pointed out that the presence of the magnetic field in the solar wind greatly reduces the mean free path perpendicular to the field lines. This reduction, in turn, reduces the lateral momentum transfer. These researches have suggested that the physical situation can be approximated by neglecting the term in $(w/r)$ in equation (3.36) and writing $T = (4/3)$ $\eta^*[dw/dr]$. They have carried this modification through and find a much more reasonable model of the solar wind with an expansion velocity at earth of about 302 km/sec.

These developments show that viscosity should be included in model solar wind calculations, but apparently care must be taken to include the viscosity and the magnetic field and perhaps other effects such as the anisotropic distribution of peculiar velocities in a consistent manner. If the interpretation suggested by Meyer and Schmidt is correct, the viscous solutions with a magnetic field resemble the inviscid solutions. Their interpretation is not universally accepted however, and the authors themselves have noted that it is probably in error. If it is incorrect, some other explanation will be needed for the serious discrepancy between the viscous solutions and the inviscid solutions, which show good agreement with the observations. The situation involves the concept and calculation of viscosity in a dilute gas which does not experience ordinary collisions.

## 3.5  Two-Fluid Models

The solar wind models described in the preceding sections of this chapter have all been, in effect, one-fluid models. The fluid was ideally composed solely of particles with mass $\mu\, m_H$, all of which can be adequately described at each point by one density, one velocity, and one temperature. This basic concept is not vitiated by evaluating the conductive heat transport with the correct expression for a proton-electron gas.

Sturrock and Hartle have pointed out that the single-fluid models contain an implicit and probably indefensible assumption concerning the rate of energy exchange between the protons and electrons. In particular, the protons and electrons could have substantially different temperatures far from the sun.

Some physical insight into the temperature inequality can be obtained by considering the expansion rate versus the energy exchange rate. The expansion rate is

$$v_{\exp} = \frac{w}{N_e}\frac{dN_e}{dr} \tag{3.37}$$

This can be approximated away from the sun where $N_e \propto r^{-2}$ by

$$v_{\exp} \approx \frac{2w}{r} \tag{3.38}$$

Near the earth, we take $w = 500$ km/sec to obtain $v_{\exp} \approx 6 \times 10^{-6}$ sec$^{-1}$. The energy exchange rate is available from standard expressions of plasma physics and is

$$v_E \approx 1 \times 10^{-1}\frac{N_e}{T_e^{3/2}} \tag{3.39}$$

Near the earth, we take $N_e = 5$/cm$^3$ and find $v \approx 5 \times 10^{-7}$ sec$^{-1}$ for $T_e = 10^4$ °K. Clearly, $T_e$ would have to be $\sim 10^3$ °K for equality of $v_E$ and $v_{\exp}$. For $v_E \gg v_{\exp}$, a $T_e$ much less than $10^3$ °K would be required. These are orders of magnitude less than the anticipated and measured values of $T_e$ throughout the interplanetary medium, and considerable differences between $T_e$ and $T_p$ would, therefore, be expected.

Hartle and Sturrock have investigated a two-fluid model of the solar wind. Only the temperatures of the two fluids can be significantly different. Both densities and flow velocities must be the same to preserve local charge neutrality and to keep the sun electrically neutral, respectively. (Recall the discussion in Section 1.2). Hence we have one equation of continuity (3.2) and one equation of motion,

$$N_e m_H w\frac{dw}{dr} = -\frac{d}{dr}[N_e k(T_e + T_p)] - \frac{GM_\odot m_H N_e}{r^2} \tag{3.40}$$

Compare equations (3.40) and (3.1). Separate energy equations are written for the electrons and protons. These are

$$\frac{3}{2} N_e wk \frac{dT_e}{dr} - wkT_e \frac{dN_e}{dr} - \frac{1}{r^2} \frac{d}{dr} \left( r^2 \kappa_e \frac{dT_e}{dr} \right)$$

$$= -\frac{3}{2} v_E N_e k(T_e - T_p) \tag{3.41}$$

and

$$\frac{3}{2} N_p wk \frac{dT_p}{dr} - wkT_p \frac{dN_p}{dr} - \frac{1}{r^2} \frac{d}{dr} \left( r^2 \kappa_p \frac{dT_p}{dr} \right)$$

$$= \frac{3}{2} v_E N_p k(T_e - T_p). \tag{3.42}$$

Compare these equations with equation (3.20). The right-hand side of these two equations represents the loss or gain of thermal energy for one fluid because of collisions with the other. Here $\kappa_e$ and $\kappa_p$ are the thermal conductivities for electrons and protons, respectively. Note again that $N_e = N_p$. Accurate expressions for $\kappa_e$, $\kappa_p$, and $v_E$ are known from plasma theory.

Hartle and Sturrock have numerically integrated this system of equations by matching their solution to the electron density in the outer corona and setting $T_e = T_p$ at the innermost point of the integration. With this model and the starting conditions just mentioned, they found at the earth, $N_e = 15/\text{cm}^3$, $w = 250$ km/sec, $T_e = 3.4 \times 10^5$ °K, and $T_p = 4.4 \times 10^3$ °K. The density is a factor two or three times higher than the observed value, and the velocity is some 50 km/sec lower than the values observed for quiet times. The proton temperature is somewhat low (although the Vela observers occasionally report values near $10^4$ °K and below, Section 5.3). The observations indicate $T_e \approx 4T_d$, and, hence, an additional energy exchange mechanism appears to be effective in the solar wind plasma.

There are also some assumptions in the two-fluid model. One is that the magnetic field and viscosity are entirely neglected. Another is that the velocity distributions of the protons and electrons are isotropic and maxwellian; this assumption involves comparison of the electron-electron relaxation rate ($v_{ee}$) and the proton-proton relaxation rate ($v_{pp}$) with the expansion rate $v_{exp}$. The numerical results fully justify the assumption anywhere within 10 a.u. Unfortunately, from observational data we know that the interplanetary magnetic field influences the temperature along the field lines such that it is approximately twice the temperature perpendicular to the field lines (see Section 5.3).

The two-fluid model also assumes no deposition of wave energy outside of $2R_\odot$. While this is probably a good first approximation, Hartle and Sturrock have attributed some of the discrepancy between observed parameters at earth and their model to additional sources of energy. Heating within the sonic critical point produces primarily an increase in velocity while heating outside the sonic critical point produces an increase in temperature.

## 3.6  Modern Exospheric Models

We will digress temporarily from our discussion of fluid models to reconsider briefly the question of exospheric models of the solar wind. As seen in Section 1.2, none of the traditional exospheric or evaporative models could reproduce the fluxes and velocities observed in the solar wind.

In exospheric theory, one seeks a critical level or base of the exosphere which is defined as the location where a particle traveling upward at speeds exceeding the escape speed has a probability $1/e$ of escaping without a collision. The escape flux is then calculated from the expression

$$F_{esc} = \frac{2\pi N_c}{U^3 \pi^{3/2}} \int_{v_{esc}}^{\infty} \int_{0}^{\pi/2} e^{(-v^2/U^2)} v^3 \cos\theta \sin\theta \, dv \qquad (3.43)$$

where $v$ is the particle velocity, $U = (2kT/m)^{1/2}$, $m$ is the mass of the escaping particle, and the escape velocity for an ionized gas is calculated from equation (1.9). Note that the escape flux is proportional to the particle density $N_c$ at the critical level.

In planetary escape calculations, a mean cross section is used to calculate one critical level, which is then used for particles with various velocities; herein lies the basic difficulty in applying these results to the solar corona. Jensen pointed out that the collision cross section for protons in a proton-electron gas varied as $v^{-4}$ for deflecting collisions (Note: the cross section varies as $v^{-6}$ for energy exchange collisions), and, therefore, the faster particles originate much deeper in the atmosphere than the mean critical level. Particles originating lower in the atmosphere come from regions with higher density and, thus, make a greater contribution to the escape flux than would be calculated from a single critical level.

Two modern exospheric models are available. Using the "cones-of-escape" approach introduced years ago by Jones, Jensen computed the contribution to the escape flux at each level in the atmosphere that resulted from collisions. Brandt and Cassinelli introduced a continuum of energy-dependent critical levels and summed over them to obtain the escape flux. Jensen's results and Brandt and Cassinelli's are in agreement. A representative corona model gives a $w = 260$ km/sec, $N_e = 3/cm^3$, and $T \approx 10^5$ °K (the temperature is computed from the velocity dispersions). These values are comparable with observed and calculated values for the solar wind, although the numerical agreement is only fair.

The principal value of the exospheric treatment is not in its numerical accuracy but in its independent production of a solar wind. Thus, the solar wind is the result of the hot solar corona and does not depend on the method of treatment.

## 3.7   Influence of the Magnetic Field

The extension of the solar magnetic field into interplanetary space is certainly important with regard to several aspects of the physics of the solar wind flow. None of the theoretical treatments discussed to this point included the field in a self-consistent manner, although Parker considered the form of the field lines when thought of as the tracer of a specific emitting region on a rotating sun. Consideration of relative energy densities immediately leads to difficulties.

The variation of the magnetic field from the sun can be calculated from

$$\mathbf{V} \cdot \mathbf{B} = 0 \tag{3.44}$$

and, hence,

$$B_r = B_\odot (r_\odot / r)^2 \tag{3.45}$$

where $B_\odot$ is the field at the solar surface. The azimuthal component can be calculated, assuming that the field is a tracer [see equation (3.15) *et sequence*]. In the plane of the solar equator,

$$B_\phi = B_r \tan \psi = B_\odot \frac{\Omega}{w} \left(\frac{r_\odot}{r}\right)^2 (r - r_\odot) \tag{3.46}$$

The total magnetic field for $r \gg r_\odot$ becomes

$$B_T = B_\odot \left(\frac{r_\odot}{r}\right)^2 \left[1 + \frac{\Omega^2 r^2}{w^2}\right]^{1/2} \tag{3.47}$$

and $B_T$ can be calculated with a model for $w(r)$.

Consider now the ratio ($\beta$) of kinetic (bulk plus thermal) energy to magnetic energy, which can be written

$$\beta = \frac{\frac{1}{2} m_H N_e w^2 + 3 N_e kT}{B_T^2 / 8\pi} \tag{3.48}$$

Equations (3.47) and (3.48) can be used to construct a table for an assumed electron density of $5/\text{cm}^3$ at earth and a field of 1 gauss at the solar surface.

The simple solar wind theory as described above certainly does not apply when $\beta \lesssim 1$. Hence on grounds of simple energetics, solar wind velocities less

TABLE 3.1.
*The Magnetic Field in a Model Solar Wind*

| $w$(km/sec) | $B_T$ (earth) ($\gamma$) | $\beta$ (earth) | $\beta$ (0.1 a.u.) |
|---|---|---|---|
| 100 | 9.5 | 1 | 0.25 |
| 300 | 4.0 | 60 | 2 |
| 1,000 | 2.5 | $2 \times 10^3$ | 25 |

than $10^2$ km/sec are suspect. The dependence of $\beta$ on $w$ for velocities of (say) 100 km/sec or less near the earth is quite steep. The thermal term would be negligible, $B_T \propto w^{-1}$, and hence, $\beta \propto w^4$. Axford, Dessler, and Gottlieb pointed out, on the basis of the argument presented above, that if $w$ is on the order of $10^2$ km/sec, the magnetic field would dominate and wind into a tight spiral; a solar wind flow would be obstructed or entirely prevented. Either the solar wind expands with sufficient energy to overcome the solar magnetic field or it does not expand at all; thus, the solar wind should have $w \gtrsim 100$ km/sec or $w = 0$. The lower limit to the solar wind speed has been observed (see Section's 4.1 and 5.2). This simplified discussion does not consider the problem of magnetic field configurations or energy ratios near the sun. Possible discussion of an interplanetary field perpendicular to the plane of the ecliptic or the solar equator is deferred until Section 5.5.

The first self-consistent discussions of a solar wind flow from a magnetized rotating sun were carried out nearly simultaneously by Weber and Davis and by Modisette. Their work is vital for discussing the solar angular momentum loss in the solar wind. All quantities are assumed to vary only with $r$ and the treatment is restricted to the plane of the solar equator. The treatment also assumes a scalar pressure, zero viscosity and $\sigma \to \infty$. The latter condition implies that the field is frozen-in [see equation (3.18) *et sequence*].

The velocity is written as

$$\mathbf{v} = w\hat{\mathbf{r}} + w_\phi\hat{\boldsymbol{\phi}} \qquad (3.49)$$

and the field as

$$\mathbf{B} = B_r\hat{\mathbf{r}} + B_\phi\hat{\boldsymbol{\phi}} \qquad (3.50)$$

where $\hat{\mathbf{r}}$ and $\hat{\boldsymbol{\phi}}$ are unit vectors in the radial and azimuthal directions, respectively. Equation (3.2) is used as the equation of continuity, but note that it may not be valid. A finite $w_\theta$ ($\theta$ being the polar angle) is permitted, but $\partial w_\theta/\partial\theta$ must be zero at the equator. Thus, no systematic flow in both hemispheres into or out of the equatorial region is permitted; such flows could be the case, however, and they should be kept in mind as a possible area of uncertainty.

The electric field in the frame of reference moving with the plasma is $\mathbf{E} + \mathbf{v} \times \mathbf{B}$; with $\sigma \to \infty$ the field must be zero to avoid large currents. Note, electromagnetic units are used throughout. Dessler has stressed the importance of the origin and conceptual basis of the ambient interplanetary electric field $E$. He pointed out that there are three equivalent ways of describing the electric field: (1) the co-rotation picture has $E = (\Omega \times \mathbf{r}) \times \mathbf{B}$. However, the interaction with the solar wind determines a geometrical relationship between the field and the plasma as shown in Figure 3.2. The magnitude of $\mathbf{E}$ is $\Omega r |\mathbf{B}| \sin \chi$. But the geometrical relationship requires that $w \cos \chi = \Omega r \sin \chi$ and that the magnitude be equal to $|\mathbf{B}| w \cos \chi = |\mathbf{B}| w \sin \psi$, a value which corresponds to $\mathbf{E} = -\mathbf{w} \times \mathbf{B}$; (2) the solar wind can be regarded as convecting a magnetic field radially away from the sun to produce a field, $\mathbf{E} = -\mathbf{w} \times \mathbf{B}$; and (3) the electric field can be described as resulting from the solar wind moving through a stationary, spiral magnetic field. Again, $\mathbf{E} = -\mathbf{w} \times \mathbf{B}$. On this picture, the field arises from a polarization of the solar wind (due to a slight charge separation); $\mathbf{E}$ is perpendicular to the plane of $\mathbf{B}$ and $\mathbf{w}$ and has magnitude $w |\mathbf{B}| \sin \psi$. Such a field produces an $\mathbf{E} \times \mathbf{B}$ drift with a velocity, $v_D = |\mathbf{E}|/|\mathbf{B}| = w \sin \psi$, and a direction perpendicular to $\mathbf{B}$. Parallel to $\mathbf{B}$, the velocity is $w \cos \psi$, and the vector sum of these two velocities is $w$. The electric field is required to force the plasma across the lines of force. Recall the phonograph record analogy of Section 3.1; only spiral field lines with the proper relationship can co-rotate. For example, kinks in the field *cannot* co-rotate and must be convected outward by the plasma. However, many quasi-stationary geometrical features can co-rotate, such as a longitudinal gradient of field magnitude.

To continue, we have

$$\mathbf{E} = -\mathbf{v} \times \mathbf{B} \tag{3.51}$$

From Maxwell's equations, we have

$$\mathbf{V} \times \mathbf{E} = -\frac{\partial \mathbf{B}}{\partial t} = 0 \tag{3.52}$$

Substituting for $\mathbf{E}$ in equation (3.52) from equation (3.51) gives

$$(\mathbf{V} \times \mathbf{E})_\phi = \frac{1}{r} \frac{d}{dr} [r(wB_\phi - w_\phi B_r)] = 0 \tag{3.53}$$

Equation (3.53) can be integrated immediately to yield

$$r(wB_\phi - w_\phi B_r) = \text{const} = -\Omega r^2 B_r \tag{3.54}$$

The constant is evaluated using the fact that $\mathbf{v} \times \mathbf{B} = 0$ in the frame rotating with the sun; this concept is nearly intuitive and was used, for example, in the derivation of the spiral angle [see the discussion following equation (3.15)]. Thus, the value of $\mathbf{v} \times \mathbf{B}$ results from the solar frame rotating at velocity $r\Omega$ cutting the radial field.

The equation of motion for the $\phi$ direction for a unit volume equates the torque to a mass times the time rate of change of angular momentum. This becomes

$$r(\mathbf{J} \times \mathbf{B})_\phi = \rho \frac{d(rw_\phi)}{dt} \tag{3.55}$$

where $\mathbf{J}$ is the current. The right-hand side of equation (3.55) is evaluated with equation (3.22) to obtain

$$(\mathbf{J} \times \mathbf{B})_\phi = \frac{\rho w}{r} \frac{d}{dr}(rw_\phi) \tag{3.56}$$

The term in $(\mathbf{J} \times \mathbf{B})$ can be evaluated from the Maxwell equation (with displacement current neglected),

$$\nabla \times \mathbf{B} = 4\pi \mathbf{J} \tag{3.57}$$

Hence,

$$(\mathbf{J} \times \mathbf{B})_\phi = \frac{1}{4\pi} [(\nabla \times \mathbf{B}) \times \mathbf{B}]_\phi \tag{3.58}$$

Evaluation of the cross products yields the $\phi$ equation of motion or

$$\frac{d}{dr}(rw_\phi) = \frac{B_r}{4\pi\rho w} \frac{d}{dr}(rB_\phi) \tag{3.59}$$

Inspection of the factor $B_r/4\pi \rho w$ shows that it is constant because both $B_r$ [equation (3.45)] and $4\pi\rho w$ [equation (3.2)] vary as $r^{-2}$. Hence, equation (3.59) can be integrated immediately to yield

$$rw_\phi - \frac{B_r}{4\pi\rho w} rB_\phi = \text{const} = L \tag{3.60}$$

where the constant $L$ is the angular momentum per unit mass. Equation (3.60) shows that angular momentum is carried not only in azimuthal particle motions but also in the azimuthal magnetic stresses. If matter escapes from the sun with a net angular momentum, then a torque is being applied to the sun and is transmitted to the surface via the magnetic stresses.

Define the radial Alfvénic Mach number as

$$M_A = \left[ \frac{4\pi\rho w^2}{B_r^2} \right]^{1/2} \tag{3.61}$$

which is just the radial solar wind speed divided by the Alfvén speed computed with the radial component of the magnetic field. Combining equations (3.61) and (3.54) gives

$$w_\phi = \Omega r \frac{(M_A^2 L r^{-2} \Omega^{-1} - 1)}{(M_A^2 - 1)} \qquad (3.62)$$

This is a very important expression. Near the sun, $M_A \ll 1$, and near the earth, $M_A > 1$. Hence, somewhere between the earth and the sun, $M_A = 1$; this point is called the Alfvén critical point with distance $r_A$ and velocity $w_A$. If $M_A = 1$, the numerator vanishes, and the denominator must vanish at $r_A$ if $w_\phi$ is to remain finite; the situation requires

$$L = \Omega r_A^2 \qquad (3.63)$$

Hence, the total angular momentum per unit mass in the solar wind can be calculated *as if* there were solid body rotation out to the point where the magnetic energy density $(B^2/8\pi)$ equals the kinetic energy in the mass flow $(\frac{1}{2} \rho w^2)$; this is the traditional result. However, the traditional result considered angular momentum in particle motions only.

The radial Alfvén speed [equation (3.61)] can be combined with equations (3.2) and (3.45) to give an estimate for $r_A$, viz.,

$$\frac{r_A}{r_E} = \frac{V_{A,E}^*}{(w_A w_E)^{1/2}} \qquad (3.64)$$

Here $V_{A,E}^*$ is the radial Alfvén velocity at earth and the subscript $E$ refers to quantities at earth. This expression can be used to estimate $r_A$ and give some idea of the rate of solar angular momentum loss. A lower limit to $r_A$ can be obtained by setting $w_A = w_E$, or

$$r_A > r_E \frac{V_{A,E}^*}{w_E} \qquad (3.65)$$

For typical parameters for the solar wind,

$$r_A \gtrsim 20 \, R_\odot \qquad (3.66)$$

Such a large value implies a large rate of angular momentum loss, which is important on time scales comparable to the age of the sun. We return to a discussion of this problem in Section 7.3.

The azimuthal velocity can be calculated from equation (3.62) using

$$M_A^2 = \frac{wr^2}{w_A^2 r_A} = \frac{\rho_A}{\rho} \qquad (3.67)$$

This produces

$$w_\phi = \frac{\Omega r}{w_A} \frac{(w_A - w)}{(1 - M_A^2)} \tag{3.68}$$

The azimuthal field can be obtained from equation (3.60), and then we have

$$B_\phi = -B_r \frac{\Omega r}{w_A} \frac{(r_A^2 - r^2)}{r_A^2(1 - M_A^2)} \tag{3.69}$$

Weber and Davis have obtained expressions for $w_\phi$ and $B_\phi$ near the sun. They also note that the relative contributions to the angular momentum loss at large distances in azimuthal fluid velocity and magnetic stresses are $[1 - (w_A/w_\infty)]$ and $[w_A/w_\infty]$, respectively.

The radial momentum equation is

$$N\mu m_H w \frac{dw}{dr} = -\frac{d}{dr}(NkT) - \frac{GN\mu m_H M_\odot}{r^2} \tag{3.70}$$

$$+ (\mathbf{J} \times \mathbf{B})_r + N\mu m_H \frac{w_\phi^2}{r}$$

where the last two terms are the radial magnetic force and the centrifugal force. The current can be obtained from Maxwell's equations (as above), and this yields

$$(\mathbf{J} \times \mathbf{B})_r = -\frac{1}{4\pi r} B_\phi \frac{d}{dr}(r B_\phi) \tag{3.71}$$

Solutions to the radial momentum equation and the equation of continuity can be obtained if an energy is specified. Weber and Davis used a polytropic law,

$$P = P_0 \left(\frac{\rho}{\rho_0}\right)^\gamma \tag{3.72}$$

where $\gamma$ is the polytropic index; they obtained the solution shown in Figure 3.6.

The "supersonic" solution is remarkable because it passes through three critical points where the flow velocity equals the velocity of a characteristic disturbance in the fluid; these are velocities corresponding to the slow mode, Alfvén mode, and fast mode in increasing order of the distance of the critical point from the sun. The slow-mode critical point is simply Parker's (sonic) critical point displaced slightly because of the magnetic field. The critical

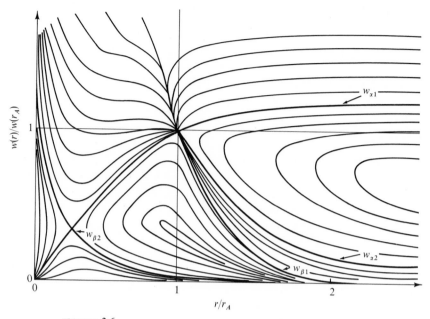

FIGURE 3.6

Solutions to the solar wind equations including the magnetic field.
(After E. J. Weber and L. Davis.)

points corresponding to the Alfvén and fast modes are very close together and cannot be distinguished on Figure 3.6.

The asymptotic behavior of the "supersonic" and several other solutions can be seen in Figure 3.6. The adopted solution $w_{\alpha 1}$ has an asymptotic velocity of 425 km/sec and a zero pressure at infinity. The solution $w_{\alpha 2}$ also has zero pressure at infinity, but the asymptotic velocity is only 9 km/sec. The solutions $w_{\beta 1}$ and $w_{\beta 2}$ have finite pressures at infinity, a behavior in contrast to Chamberlain's solar breeze solutions (without magnetic field), which had zero pressure at infinity. Since zero pressure at infinity is undoubtedly a requirement for acceptable solutions, inclusion of the magnetic field appears to rule out the solar breeze solutions. This conclusion was implied in the discussion of the theories of Axford, Dessler, and Gottlieb earlier in this section.

The solution obtained by Weber and Davis was close to observed solar wind conditions near the earth (chosen as their starting point) and clearly indicated the topology of the situation; the azimuthal velocity was 1 km/sec. However, there was poor agreement with coronal conditions: the temperature was too high and the density very low. This situation prompted Brandt, Wolff, and Cassinelli to reinvestigate the problem from the numerical point of view with the full energy equation instead of equation (3.72).

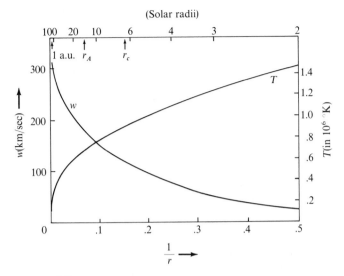

FIGURE 3.7
The run of $T$ and $w_r$ on the solar wind model discussed in the text.
(After J. C. Brandt, C. Wolff, and J. P. Cassinelli.)

The energy equation for this problem is

$$
F_E = 4\pi\mu m_H C \left[ \frac{-\kappa_0 T^{5/2}}{m_H \mu N_e w} \frac{dT}{dr} (\cos \theta) \right.
$$

$$
+ \frac{5kT}{\mu m_H} + w^2 + w_\phi^2 - \frac{2GM_\odot}{r}
$$

$$
\left. - \frac{B_\phi B_r \Omega r}{4\pi N_e m_H \mu w} \right] + 4\pi r^2 [F_R + b]
$$

(3.73)

The contributions to the energy flux in the right-hand side are, in order: conduction, reduced by $\cos \theta$ because of the reduction in conductivity perpendicular to the field lines ($\theta$ being the angle between the field lines and the radius vector); thermal energy plus the potential energy for adiabatic expansion; kinetic energy (radial); kinetic energy (azimuthal); gravitational potential energy; magnetic energy; photon energy as a result of radiative losses; and mechanical energy ($b$). The last two terms in the energy flux are probably negligible except in the very low corona. The equation used in the solution was for $(F_R + b) = 0$, and the solution was fit to coronal electron densities for $r > 2R_\odot$. In addition, the numerical solution requires $T \to 0$ as $r \to \infty$.

In principle, the numerical solution is not difficult because Weber and Davis delineated the topology of the solution. The adopted solution near earth is shown in Figures 3.7 and 3.8, and it has $N_e = 6/\text{cm}^3$, $w = 315$

88

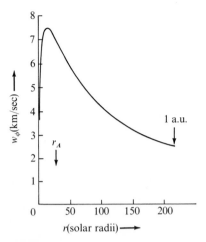

FIGURE 3.8
The variation of $w_\phi$ on the solar wind model
discussed in the text.
(After J. C. Brandt, C. Wolff, and J. P.
Cassinelli.)

km/sec, $T = 325{,}000\ °K$, a spiral angle of $55°$, and $w_\phi = 2.5$ km/sec. Such values are reasonable for a quiet solar wind.

Nevertheless, the value found for $w_\phi$ could be somewhat low, and larger values of $w_\phi$ could be indicated for average or disturbed conditions. One explanation may be that there are departures from azimuthal symmetry. Another explanation has been proposed by Schubert and Coleman, who pointed out that angular momentum could be carried in the density fluctuations associated with magnetosonic waves. Rough estimates indicate that this process could be important particularly in connection with the higher values of $w_\phi$ thought to occur at times of substantial geomagnetic activity.

## 3.8   Nonuniform Flow

The models considered in the preceding sections of this chapter have been severely idealized as being time independent and having spherical or azimuthal symmetry. Models without these assumptions are extremely difficult to obtain and quantitative statements are difficult to make with confidence. There appears to be a plethora of phenomena associated with nonuniform flow and waves and/or discontinuities (Section 3.9) in the solar wind. Full understanding is yet to come. Here we will discuss briefly both the physics of enhanced flow from a local solar region and the physics of a "blast wave" propagating into the solar wind after a solar flare.

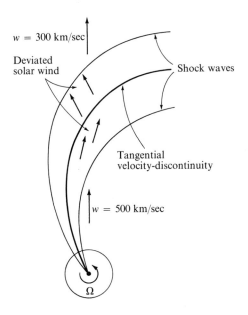

FIGURE 3.9
Schematic of the formation of a contact surface in the
solar wind because of nonuniform flow. (After A. J.
Dessler.)

Consider first the variation of solar wind speed with solar longitude; suppose that a region producing a 300 km/sec solar wind precedes (in the sense of solar rotation) a region producing a 500 km/sec solar wind. As the sun rotates, the slow-moving plasma is exposed to the faster beam, and a collision of the two plasmas occurs at a relative velocity of 200 km/sec; and in a quasi-stationary situation, this "collision" co-rotates.

The magneto-acoustic velocity in the solar wind near the earth is close to the Alfvén speed [equation (2.6)], which is about 70 km/sec. For collisions at velocities greater than the magneto-acoustic velocity, a double shock is formed that is separated by a tangential velocity discontinuity as shown in Figure 3.9. The flow is deviated against the solar rotation between the inner shock and the discontinuity; it is deviated with the solar rotation between the discontinuity and the outer shock. These changes in flow direction produce the tangential velocity-discontinuity. If the relative velocities are smaller than the magneto-acoustic velocity, a similar picture develops without the shocks.

The stability of this situation has been considered, and apparently such a "feature" could persist for several rotations. However, the region between the shocks is unstable, and a region of disordered fields should develop. Such a co-rotating feature could be the source of the geomagnetic disturbances

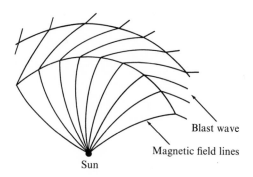

FIGURE 3.10

The magnetic field structure in an interplanetary
blast wave. (After E. N. Parker.)

ascribed to the solar M-regions. The physical origin would be a longitudinal
velocity gradient in the solar wind, and there might not be an association with
features on the solar surface. The general physics involved could be important
in the discussion of the sector structure, and we return to this subject in
Section 5.6.

Flare-produced effects in the solar wind are important for the discussion of
cosmic ray events, such as the Forbush decrease following a flare. Parker has
shown that there are no stationary solutions if a portion of the corona is
heated by a flare event to a temperature exceeding a certain value. He argues
that the result of such an "explosion" would be a blast wave propagating into
the interplanetary medium. The wave is composed of the quiet solar wind,
swept up by the momentum of the coronal explosion, and the coronal gases
themselves at the rear of the wave. Parker has worked out the details for
several possible cases, and a sample is shown in Figure 3.10. A shock forms at
the head of the blast wave; it can be considered as a bow wave since the blast
wave acts like a piston in the interplanetary medium. The magnetic field in the
quiet flow is sheared in the shock front, and the cosmic rays are swept out by
the compressed field to produce the Forbush decrease. The bow wave appears
to produce the initial sudden storm commencement (SSC) in geomagnetic
storms while the higher densities towards the rear of the blast wave produce
the main phase of the storm. The standoff distance of the shock is approximately proportional to the radius of curvature of the gas acting as the "piston."
Parker has estimated that these blast wave events can make an important
contribution to the mass loss in the solar wind. In addition, these blast waves
can be dissipated at large distances from the sun to produce additional
expansion or an increase in temperature.

An alternate model of the geomagnetic event has been presented by Gold
and others. They postulated that magnetic tongues or loops of the re-entrant

field are carried out of the sun by the energy imparted from the flare event This tongue envelopes the earth and shields it from cosmic rays to produce the Forbush decrease. It is difficult to imagine how this event could occur without first having the event described above as advocated by Parker; moreover, such a model implies fairly frequent reversal of signs in the interplanetary magnetic field, which have not been observed. At present, this theory has not been generally accepted. However, the general type of event envisioned by Gold appears to occur in the formation of new sectors in the interplanetary medium (see Section 5.6.)

## 3.9 Plasma Instabilities and Waves

The possibility of tangled or disordered magnetic fields in the interplanetary medium was first inferred from evidence concerning the storage in the solar system of cosmic rays from solar flares (see Section 6.7). Parker has considered some possible mechanisms for the production of disordered fields through plasma instabilities that are caused by pressure anisotropies.

One such is the fire-hose instability, which occurs when the pressure perpendicular to the magnetic field $(P_\perp)$ is less than the pressure parallel to the field $(P_\parallel)$ according to the relation

$$P_\parallel - P_\perp > B^2/4\pi \tag{3.74}$$

This is the condition for instability. If the field is essentially radial and the flow is also radial and at constant velocity, the radial thermal velocities remain approximately constant (no expansion in the radial direction) while the perpendicular thermal velocities decrease adiabatically because of the radial divergence. Thus, a tendency toward $P_\parallel > P_\perp$ develops, and this is counteracted by collisions and the plasma instabilities under discussion.

Equation (3.74) can be interpreted as follows. Consider a perturbation in the field with radius of curvature R. The particles moving along the field line experience a centripetal force $\rho v_\parallel{}^2/R \approx P_\parallel/R$, which makes the perturbation grow. The restoring forces are the gradient in $P_\perp$ which is $\approx P_\parallel/R$ and $\approx B^2/4\pi R$ (resulting from the tension in the field lines). If the restoring forces do not exceed $P_\parallel/R$, we have equation (3.74). In essence, the concentration of particle motions along the field lines converts the tension in the field lines $(B^2/4\pi)$ into a compression with the resultant buckling of the field lines.

The characteristic growth time for disturbances with wavelength $10^{-2}$ a.u. is about four hours. The observations of the solar wind plasma near earth indicate $P_\parallel \approx 2P_\perp$, and, hence, the instability criterion becomes $4\pi N_e k T_\perp > B^2$. For $N_e = 5/\text{cm}^3$, $T = 10^5$ °K, and $B = 5\gamma$, the instability criterion is not satisfied. It is probably not satisfied very often near the orbit of the earth.

Another instability is the mirror instability, which could be important in the solar wind beyond the earth. The instability criterion is

$$\frac{P_\perp}{P_\parallel} > P_\perp + \frac{B^2}{8\pi} \tag{3.75}$$

As expected from equation (3.17), the field lines beyond the earth wind tightly and become essentially azimuthal; for example, the spiral angle on the solar wind model cited in Section 3.7 is 55° at 1 a.u. and increases to 80° at 5 a.u. Then, if the radial pressure is larger than the pressure perpendicular to the radial direction, equation (3.75) could be satisfied with $P_\perp > P_\parallel$ (where $\perp$ and $\parallel$ refer to the magnetic field). If $P_\perp > P_\parallel$, then most pitch angles will be large; a perturbation as shown in Figure 3.11 produces magnetic mirrors (by the first magnetic invariant; see equation 5.2), and the particles with large pitch angles are reflected away from the regions of larger field to the regions of smaller field. Thus, the ions tend to expand the regions of weak field, which become weaker. Growth rates for the mirror instability are usually comparable with growth rates for the fire-hose instability for comparable wavelength and degree of anisotropy. The mirror instability is thought to be important beyond the orbit of the earth and may provide the irregularities required for solar cosmic ray storage.

However, fluctuations in the solar wind magnetic field near the earth could be interpreted as magneto acoustic waves; these could originate at the sun or from the interaction of fast and slow solar beams, as discussed above. The tangential velocity discontinuities are, in effect, the boundary between two fluids in relative motion parallel to the boundary and, hence, subject to the Kelvin-Helmholtz instability. This instability could be the mechanism for wave production from the many discontinuities found observationally; see also Section 5.6. Consider a spherical hydromagnetic wave propagating radially with no dissipation. Conservation of energy implies

$$(\delta B)^2 \, (v_H + w) \, r^2 = \text{const} \tag{3.76}$$

where $\delta B$ is the amplitude of the wave, $v_H$ is the appropriate hydromagnetic velocity, $w$ is the solar wind speed, and $r$ is the distance. This equation is analogous to equation (2.5). Since $B$ varies approximately as $r^{-2}$, we can write

$$\frac{\delta B}{B} \propto \frac{r}{(v_H + w)^{1/2}} \tag{3.77}$$

At a great distance from the sun, $w$ is constant and much greater than $v_H$; hence, one finds $\delta B/B \propto r$ (valid for a predominantly radial field). The point is that hydromagnetic waves can grow significantly in amplitude as they are

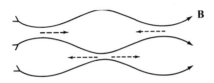

FIGURE 3.11
The field configuration leading
to the mirror instability.

convected outward by the solar wind. If $\delta B$ becomes $\approx B$, the waves could develop into shocks and provide a source for energy deposition far from the sun.

Scarf found the fastest growing waves to be in the magnetosonic mode with frequencies near the ion cyclotron frequency. Such resonant magnetosonic modes may be very important in any consideration of the basic physics of the solar wind as a mechanism providing coupling between particles. Today, the trend in experiments from the observational or experimental viewpoint, is toward the understanding of the basic physics of the solar wind plasma.

In summary, one could draw the following conclusions: (1) instabilities or waves seem significant enough to guarantee the presence of irregularities in the interplanetary medium (as observed); (2) these waves and instabilities serve to produce "effective collisions" between particles, help ensure the fluid behavior of the solar wind, and tend to equalize the proton and electron temperatures. The instabilities also keep the thermal anisotropies from reaching extreme values; and (3) these considerations are largely qualitative and are rarely included in quantitative models. They serve as an additional warning against taking the idealized models too literally. Observationally, directional discontinuities in the solar wind seem very common (Section 5.6); these could arise from velocity gradients in the solar wind, as outlined in Section 3.8.

## 3.10 Termination of the Solar Wind

Here we discuss the extension of the solar wind into the local interstellar medium and the nature of the interaction involved. The energetic particles could ionize any neutral hydrogen in the local interstellar gas and thus the solar wind could be instrumental in determining the size of the solar H II region or "Strömgren sphere."

The interstellar medium seems to present an obstacle to the supersonic or super-Alfvénic solar wind. Clauser and Weymann have pointed out that a shock transition should occur between the solar wind and the interstellar medium. A supersonic wind cannot receive the information that an obstacle exists ahead; the shock transition from supersonic to subsonic flow permits the information to be transmitted and allows a smooth flow around or into an "obstacle."

We find in the subsequent discussion that the interstellar magnetic field actually presents the "obstacle" to the solar wind. With this fact in mind, refer to Figure 3.12. The distance $r_s$ is the distance to the shock transition; at this point, the velocity decreases and the temperature increases to a value comparable with the energy of the directed solar wind motion. The distance $r_m$ is the distance to the pushed-back interstellar magnetic field; the topology of this interface could be quite complex. Some penetration of shocked solar wind material into the interstellar medium is expected via plasma instabilities.

The actual state of the local interstellar medium is unknown. The mean density of the gas near the sun (averaged over a large volume) is generally considered to be 1 hydrogen atom/cm³, either neutral or ionized. This average is probably composed of a background density of 0.1/cm³ (neutral or ionized) with the neutral hydrogen concentrated in discrete clouds with densities $\sim 10/\text{cm}^3$. If the local interstellar gas is neutral, $r_I$ denotes the distance where the degree of ionization,

$$x = \frac{N_e}{N_H + N_e} \tag{3.78}$$

is approximately one-half. If, on the other hand, the local interstellar gas is kept ionized by the ultraviolet radiation ($\lambda < 912$ Å) of relatively nearby stars with high effective temperatures (such as $\gamma$ Velorum and $\zeta$ Puppis), the concept of $r_I$ is meaningless. The temperature of the solar wind must ultimately decrease to the interstellar value of $\sim 10^2$ °K (for a neutral interstellar gas) or $\sim 10^4$ °K (for ionized interstellar gas).

The solar wind near the earth has an approximately constant (with $r$) velocity of 400 km/sec and an electron density of 5/cm³ (decreasing as $r^{-2}$). This flow continues until the energy density of the flow or the impact pressure is comparable with the interstellar pressure, $P_i$. Thus, we seek the point at which

$$\tfrac{1}{2} m_H w^2 N_e = P_i \tag{3.79}$$

Since we can write

$$N_e = N_E \left(\frac{r_E}{r}\right)^2 \tag{3.80}$$

where $N_E$ is the electron density at earth, equation (3.79) becomes

$$r_s = r_E \left[\frac{N_E w^2 m_H}{2 P_i}\right]^{1/2} \tag{3.81}$$

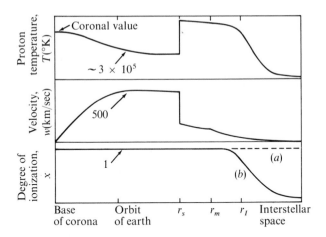

FIGURE 3.12
Schematic of the distances and conceptual points
involved in a model of the solar wind. Curves
marked (*a*) and (*b*) refer to the cases of ionized
and neutral local interstellar medium,
respectively. (After J. C. Brandt from *Icarus.*)

and $r_s$ can be calculated from measured solar wind parameters and an estimate
of the relevant interstellar pressure.

Interstellar pressure is considered to be made up of three pressures: gas
pressure, pressure from the interstellar magnetic field, and cosmic ray pressure.
For a density of one hydrogen/cm$^3$ and the temperatures described above, the
gas pressure is $2.8 \times 10^{-12}$ dynes/cm$^2$ and $1.4 \times 10^{-14}$ dynes/cm$^2$ for the
ionized and neutral state, respectively; these values are a factor of ten lower
for a hydrogen density of 0.1/cm$^3$. The magnetic field pressure $(B^2/8\pi)$ is
$4 \times 10^{-12}$ dynes/cm$^2$ for a field of $1\gamma$. The cosmic ray pressure should be
approximately equal to the pressure of the magnetic field. Only a fraction of
the cosmic ray pressure acts on the solar wind because some of the cosmic
rays diffuse into the solar system. This could be partially balanced if some of the
solar wind particles penetrate the interstellar magnetic field via plasma instabilities; a boundary between plasma and field is unstable and subject to the
interchange instability if the radius of curvature of the boundary lies in the
plasma.

The physical origin of this instability can be seen by consideration of a pure
plasma—pure magnetic field interface for the case of concave and convex
boundaries. For the boundary concave toward the plasma, the field increases
to a maximum at the boundary, and the magnetic and kinetic pressure are equal

there. If a wave-like perturbation to the boundary is applied in such a way that the volume of the plasma, and hence the pressure, remains constant, the troughs in the plasma will have a field pressure greater than the kinetic pressure and the crests in the field will have a kinetic pressure greater than the magnetic field pressure. The effects of these imbalances is to increase the perturbation and the situation is unstable. This is the case for the solar wind-interstellar field boundary. For the boundary convex toward the plasma, the field decreases to a minimum at the boundary and by a similar argument leads to the reverse or stable situation. This is the case for the solar wind-magnetosphere boundary (Section 6.3).

It is probably not too far wrong to assume a total interstellar pressure of about $8 \times 10^{-12}$ dynes. This value and the solar wind parameters quoted above can be substituted into equation (3.81) to obtain $r_s \approx 30$ a.u. If the cosmic ray pressure is neglected, $r_s \approx 40$ a.u. Note that the energy in the azimuthal component of the magnetic field has been entirely neglected in the discussion of the deviation of the solar wind flow.

Conservation of matter and energy at the shock lead to the Rankine-Hugoniot relations; for the appropriate case of a strong shock (high Mach number), these relations become

$$w_2 = w_1 \frac{\gamma - 1}{\gamma + 1} \tag{3.82}$$

$$N_2 = N_1 \frac{\gamma + 1}{\gamma - 1} \tag{3.83}$$

and

$$P_2 = \frac{2}{\gamma + 1} N_1 m_H w_1^2 \tag{3.84}$$

Here subscript one refers to in front of the shock and subscript two to behind the shock; $\gamma$ is the ratio of specific heats which is 5/3 for a plasma. Thus, the velocity drops by a factor of four (from 400 km/sec to 100 km/sec), and the density increases by the same factor. Equations (3.83) and (3.84) can be combined with the perfect gas law to compute the temperature of the shocked gas, viz.,

$$T_2 = \frac{3}{16} \frac{m_H w_1^2}{k} \tag{3.85}$$

If the impact pressure is distributed to both the electrons and protons, the initial velocity of 400 km/sec produces a shocked gas with a temperature of $3.5 \times 10^6$ °K. If only the protons are thermalized, the temperature would be a factor of two higher.

The discussion to this point has assumed a spherically symmetric situation. The solar wind emission itself is far from being spherically symmetric. If the interstellar magnetic field stops the solar wind, the flow between the shock and the pushed-back field is likely to be channeled into the direction parallel to the local interstellar magnetic field. In addition, the local interstellar gas can exert an asymmetrical impact pressure on the solar wind by virtue of the sun's peculiar motion of 20 km/sec toward an apex at $\alpha$ (right ascension) $\approx 18^h$ and $\delta$ (declination) $\approx +30°$. Thus, there are several likely sources of asymmetry in the distant solar wind.

The injection of a hot gas from the solar system into the local interstellar gas has some interesting consequences. Patterson, Johnson, and Hanson pointed out that the energetic protons could charge-exchange with any available neutral hydrogen atoms (presumably of interstellar origin) to produce hot neutral hydrogen atoms. Some of these could penetrate back into the solar system near the orbit of earth to produce a small but significant population of energetic neutral hydrogen in the interplanetary medium. Neutral hydrogen with velocities of a few hundred km/sec can penetrate well into the inner solar system before being photoionized or charge-exchanged. It has been assumed that the creation of hot neutral hydrogen takes place in a thin shell located just beyond $r_s$ at about 20 a.u. The densities were thought to be sufficient to produce the part of the night sky Lyman-$\alpha$, which appears to come from a high temperature source. Hundhausen has investigated some of the details of this picture and found that the maximum production of hot neutral hydrogen atoms occurs in a shell about 10 a.u. thick at a distance of about $4r_s$. In this case, the shock would have to be at about 5 a.u., a value which conflicts with the much larger value of 30–40 a.u. found from pressure balance.

Part of the discrepancy may arise from the assumption of appreciable interstellar neutral hydrogen near the sun. If the local interstellar medium were ionized, the "hot" component of the night sky Lyman-$\alpha$ radiation could be diffuse intergalactic Lyman-$\alpha$, or it could originate in the earth's atmosphere. Hydrogen near the sun can be ionized by other stars (as noted above), but the solar wind itself also has an ionizing influence.

Consider the hypothesis that the ionized region around the sun is in a background of neutral galactic hydrogen. Then the solar ultra-violet radiation would not be effective in ionizing neutral hydrogen, and neutral hydrogen with a velocity of 20 km/sec (due to the sun's motion) could approach to within 10 a.u. before ionization by solar radiation. Ionization by solar wind particles can be roughly treated by introducing an ionization velocity, $V_I$. Suppose the sun is placed in an infinite homogeneous medium with hydrogen density $N_H$. If recombination is negligible and the ionization occurs near the boundary, the boundary will expand with velocity $V_I$ defined by

$$4\pi r^2 N_H V_I = J_e Y_e \tag{3.86}$$

where $Y_e$ is the total efflux of electrons from the sun and $J_e$ is the total average number of electrons produced by each electron. Thus, the boundary moves outward at the speed set by the ionization rate.

In the case of the solar wind, hydrogen is supplied to it by the sun's peculiar motion; if the hydrogen diffused radially inward at velocity $V_H$, the boundary of the solar ionized region would be where $V_I = V_H$; this gives

$$r_I^{(1)} = \left[ \frac{J_e Y_e}{4\pi N_H V_H} \right]^{1/2} \tag{3.87}$$

Note that this boundary is not sharp and specifically not the same as the classical " Strömgren sphere " where the stellar radiation is severely attenuated by absorption in the Lyman continuum; here, the region is optically thin to radiation. Also note that $r_I^{(1)}$ is strictly a characteristic size. Typical flux values for the solar wind give a total efflux of about $1 \times 10^{36}$ electrons (or protons) per second. If the electrons are thermalized beyond the shock, then energy would be about 0.5 kev. Electrons with such energies can ionize about fifteen times; hence, $Y_e J_e \approx 15 \times 10^{36}$ sec. The inbound motion is due entirely to the solar motion of 20 km/sec. Neutral hydrogen is probably not attracted into the solar system by the solar gravitation because the radiation pressure of the solar Lyman-$\alpha$ emission line almost exactly balances solar gravity. Radio astronomy studies show that systematic motions of neutral hydrogen near the sun are much less than the sun's peculiar velocity. The quantity $V_H$ in equation (3.87) refers to a radial flow. Here the neutral hydrogen impinges on only one hemisphere, and the mean velocity over the hemisphere is one-half the directed value; hence the effective value of $V_H$ is $(1/4) \times 20$ km/sec $\approx 5$ km/sec. These numbers and $N_H = 0.1/cm^3$ yield

$$r_I^{(1)} \approx 300 \text{ a.u.} \tag{3.88}$$

The variation in this number with different parameters can be seen from equation (3.87). Again note that $r_I^{(1)}$ is a characteristic radius.

If the directed energy of the solar wind is retained entirely by the protons, the energy of the electrons would be small and little ionization would take place. In this case, consideration of a momentum balance is in order; the collision cross sections are high enough to couple together the outflowing ionized gas and the inflowing neutral hydrogen. The momentum balance depends on the fraction of momentum $R$ that is transmitted into the confining magnetic region by magnetic instabilities. For typical solar wind parameters and $R \sim 0.10$, an ionized region maintained by the sweeping action of the solar protons is found with $r_I^{(2)} \approx 300$ a.u. The region is asymmetrical with a minimum dimension in the direction of the sun's peculiar motion (apex); the region is essentially unbounded in the antapex direction. The solar plasma flows out of this region before recombination can take place. Hence, the action of the solar wind is probably sufficient to maintain an ionized region with a characteristic radius of hundreds of a.u.

If these figures are correct, it is unlikely that significant numbers of fast neutral hydrogen atoms in the inner solar system could originate by way of charge-exchanges between neutral hydrogen and solar wind protons. However our knowledge of the distant solar wind is meagre, and measurements of the density of fast neutral hydrogen atoms near the orbit of the earth (but free of terrestrial influences) could provide our first direct information on the solar wind far beyond the earth. Evidence pertaining to the discovery of $He^+$ in the solar wind (Section 5.4) could argue *against* the picture described here and *for* the presence of fast hydrogen atoms in the interplanetary medium.

In summary, the supersonic solar wind velocities are terminated at a few tens of a.u. by an interaction with the interstellar magnetic field; the termination takes the form of a shock transition to subsonic flow. Some of this hot gas is able to penetrate the concave unstable boundary of the interstellar magnetic field to produce an ionized region that could, if neutral galactic hydrogen exists locally, extend to some hundreds of a.u.

# Bibliographical Notes: Chapter 3

*General References*
Some review articles or books dealing with all or a large portion of solar wind theory follow:
1. Parker, E. N.: *Interplanetary Dynamical Processes*, Interscience Publishers, New York (1963).
2. Parker, E. N.: *Space Sci. Revs.*, **4**, 666 (1965).
3. Dessler, A. J.: *Revs. Geophys.* **5**, 1 (1967).

*Section* 3.1
See the general reviews cited and:
4. Parker, E. N.: *Ap. J.*, **128**, 664 (1958).

The designation "solar wind" was first applied in:
5. Parker, E. N.: *Phys. Fluids*, **1**, 171 (1958).

Some difficulties with the high fluxes found on Parker's early models are illustrated in:
6. Newkirk, G. A., J. W. Warwick, and H. Zirin: *J. Geophys. Res.*, **65**, 2540 (1960).

*Section* 3.2
See the general review references and:
7. Chamberlain, J. W.: *Ap. J.*, **133**, 675 (1961).
8. Chamberlain, J. W.: *Ap. J.*, **141**, 320 (1965).
9. Parker, E. N.: *Ap. J.*, **141**, 322 (1965).
10. Dahlberg, E.: *Ap. J.*, **140**, 268 (1964).
11. Carovillano, R. L., and J. H. King: *Ap. J.*, **141**, 526 (1965).
12. Noble, L. M., and F. L. Scarf: *Ap. J.*, **138**, 1169 (1963).
13. Scarf, F. L., and L. M. Noble: *Ap. J.*, **141**, 1479 (1965). (Note the numerical correction to Ref. 12.)
14. Whang, Y. C., and C. C. Chang: *J. Geophys. Res.*, **70**, 4175 (1965).
15. Kopp, R. A.: Thesis Astronomy Department, Harvard University (1966).
16. Jockers, K.: *Solar Phys.*, **3**, 603 (1968).

*Section* 3.3
See Reference 3 and additional references given there.

*Section* 3.4
The problem of the viscous solutions is discussed in References 2, 11, 12, and:
17. Whang, Y. C., C. K. Liu, and C. C. Chang: *Ap. J.*, **145**, 255 (1966).
18. Meyer, F., and H. U. Schmidt: *Mitt. d. Astron. Gesell.*, **21**, 96 (1966), and **25**, 228 (1968).

*Section* 3.5

The two fluid models discussed in the text are from:

19. Hartle, R. E., and P. A. Sturrock: *Ap. J.*, **151**, 1155 (1968).

*Section* 3.6

The exospheric models mentioned are found in:

20. Jensen, E.: *Astrophys. Norvegica*, **8**, 99 (1963).

21. Brandt, J. C., and J. P. Cassinelli: *Icarus*, **5**, 47 (1966).

*Section* 3.7

The magnetic field is discussed in the general references and in:

22. Axford, W. I., A. J. Dessler, and B. Gottlieb: *Ap. J.*, **137**, 1268 (1963).

23. Weber, E. J., and L. Davis: *Ap. J.*, **148**, 217 (1967).

24. Modisette, J. L.: *J. Geophys. Res.*, **72**, 1521 (1967).

25. Weber, E. J. and L. Davis: *Trans. Am. Geophys. Union*, **48**, 171 (1967).

26. Brandt, J. C., C. Wolff, and J. P. Cassinelli: *Ap. J.*, **156**, 1117 (1969).

27. Schubert, G., and P. J. Coleman: *Ap. J.*, **153**, 943 (1968).

*Section* 3.8

Nonuniform flow is discussed in References 1, 2, 3, and thoroughly reviewed in:

28. Colburn, D. S., and C. P. Sonett: *Space Sci. Revs.*, **5**, 439 (1966).

Also consult the references to Section 5.6.

*Section* 3.9

Waves and instabilities are discussed in References 1, 2, 3, and:

29. Scarf, F. L., J. H. Wolfe, and R. W. Silva: *J. Geophys. Res.*, **72**, 493 (1967).

The standard reference for instabilities is:

30. Chandrasekhar, S.: *Hydrodynamic and Hydromagnetic Stability*, Clarendon Press, Oxford (1961).

See also the references to Section 5.6.

*Section* 3.10

The subject of the extension of the solar wind into space is treated in the general review references, Reference 22, and in:

31. Davis, L.: *Proc. Intern. Conf. Cosmic Rays and Earth Storm, J. Phys. Soc. Japan*, **17**, Suppl. A-II, 543 (1962).

32. Patterson, T. N. L., F. S. Johnson, and W. B. Hanson: *Planetary Space Sci.*, **11**, 767 (1963).

33. Brandt, J. C.: *Icarus*, **3**, 253 (1964).

34. Hundhausen, A. J.: *Planetary Space Sci.*, **16**, 783 (1968).

# 4

# *Ground-Based Methods of Observation*

A great deal of qualitative and quantitative information about the solar wind can be obtained from studies which do not require deep-space probes. Historically, these studies provided information and impetus before the advent of space observations, and they continue to provide information about times and locations (e.g., out of the plane of the ecliptic) currently inaccessible to space probes. The major studies have been of ionic comet tails, radar reflections, solar noise bursts and other radio observations, geomagnetic observations, and cosmic rays.

## 4.1   Ionic Comet Tails

The study of ionic comet tails has contributed a great deal to the history and present knowledge of the solar wind. The early studies of comet knots and their motion by Biermann led directly to many of our modern concepts of the solar wind (see the discussion in Section 1.3). Biermann also carried out the first studies of the structure of ionic comet tails; such studies have not been exhausted, particularly in the area of associations between solar-geomagnetic activity and cometary activity as evidenced by the number and brightness of structures (kinks, knots) seen in $CO^+$. In another area, ionic comet tails observed at high solar latitudes appear entirely normal. This leads to the conclusion that the polar solar wind is qualitatively similar to the equatorial solar wind. Similarly, the existence of an ionic comet tail can be taken as an indicator of the solar wind. Since ionic comet tails are regularly observed out to heliocentric distances of about 2 a.u., we may conclude that the solar wind normally extends at least that far. One extraordinary comet (Humason, 1961e) showed $CO^+$ emission out to 5–6 a.u., which, presumably, indicates a solar wind at this distance—at least at the time of observation. However, the appearance of this comet is quite unusual, and its classification as a " normal Type I " comet is doubtful.

The study of comets has also led to the detection of large-scale irregularities, such as the velocity discontinuities mentioned in Section 3.8. A beautiful example was found on July 4, 1964, by Biermann and Rh. Lüst for Comet Tomita-Gerber-Honda (1964c), which showed a tail about 30° in length with a sharp, pronounced kink at some 24° from the nucleus. The kink was nearly perpendicular to the radius vector and gave the appearance of having been swept up by a front of fast-moving plasma. This front was traced to a very long-lived (lifetime of years) source of geomagnetic disturbances.

The most interesting and profitable aspects of comet studies that pertain to the properties of the solar wind are those which investigate the orientations of the ionic or Type I comet tails (Figure 4.1). (The historical work by Hoff-meister was presented in some detail in Section 1.3.) Several compilations of tail orientation data have been made, and the most extensive is the comet catalogue of Belton and Brandt. Many of the techniques that were used in the catalogue were the result of the pioneering study by Osterbrock of Comet Baade (1954h) and Comet Haro-Chavira (1954k), which stimulated consider-able interest in the orientations of comet tails.

This catalogue contains approximately 1,600 observations (both original and previously published ones). The prime data is the position angle of the tail $\theta$ on the plane of the sky (measured east from north). The geometry on the plane of the sky is shown in Figure 4.2; the vectors are $\mathbf{r}$ (the prolonged radius vector), $\mathbf{t}$ (the tail vector), and $-\mathbf{V}$ (the negative velocity vector of the comet). The position angles of $\mathbf{r}$ and $-\mathbf{V}$ are determined from the position of the comet and the knowledge of the comet's orbit.

The plane of the comet's orbit is a convenient reference plane for the inter-pretation of comet tail orientations. The quantities $\mathbf{r}$ and $\mathbf{V}$ lie in this plane, and we can calculate the position of the tail by assuming that it lies in the orbital plane also. This situation is shown in Figure 4.3; $\gamma$ is the angle between the prolonged radius vector and the comet's velocity vector, and $\varepsilon$ is the aberration angle of the tail with respect to the radius vector. Many other quantities relating to geometrical circumstances, etc., are tabulated for each observation in the Belton and Brandt catalogue.

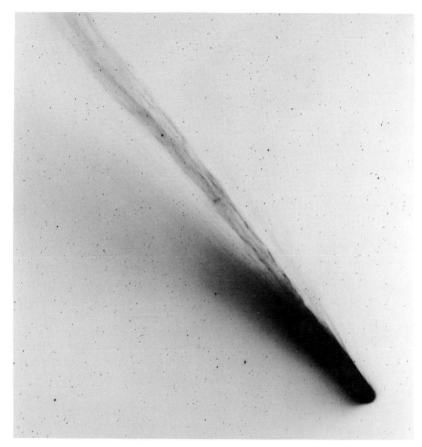

FIGURE 4.1
A photograph of Comet Mrkos clearly showing the straight filamentary Type I tail
($CO^+$) and the curved homogeneous Type II tail (dust). (Courtesy Mount Wilson and
Palomar Observatories.)

The basic equation for the interpretation of the comet observations is

$$\mathbf{T} = \mathbf{w} - \mathbf{V} \tag{4.1}$$

where $\mathbf{T}$ is the tail vector. It is distinguished from the vector $\mathbf{t}$ (which lies in
the comet's orbital plane) to avoid any prior assumptions concerning its
location (see Figures 4.2 and 4.3). Equation (4.1) indicates that the tail
points in the direction of the solar wind as seen by an observer riding on the
comet.

If the comet is near the solar equator and we idealize $w$ as having the components radial ($w_r$) and azimuthal ($w_\phi$), all quantities can be projected into
the comet's orbital plane to obtain

$$\tan \varepsilon = \frac{V \sin \gamma - w_\phi \cos i'}{w_r - V \cos \gamma} \tag{4.2}$$

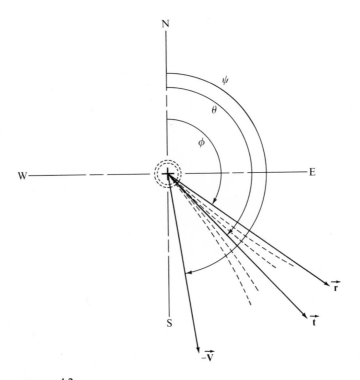

FIGURE 4.2
The geometry of a comet tail observation as seen on the plane of the sky. The position angles are: $\phi$, the prolonged radius vector; $\theta$, the comet tail; $\psi$, the direction of the comet's velocity back along the orbit. (M. J. S. Belton and J. C. Brandt from the *Astrophysical Journal Supplement*, No. 117, Copyright 1966 by The University of Chicago.)

Here $i'$ is the inclination of the comet's orbit to the plane of the solar equator. Note that $\cos i'$, $V$, $\gamma$, and $\tan \varepsilon$ are measured or calculated. The unknowns are $w_r$ and $w_\phi$; since $w_\phi$ is small, a value of $w_r$ can be approximately calculated from equation (4.2).

This calculation has been done directly from equation (4.2) for the observations with large $\varepsilon$ where the errors of measurement are relatively small compared with $\varepsilon$. The errors are calculated from known errors of measurement of the position angle $\theta$ and $d\varepsilon/d\theta$ calculated from the geometrical circumstances. This calculation of solar wind velocities for the larger values of $\varepsilon$ directly determines the smaller values of $w_r$ [see equation (4.2)]. When these results are plotted in a histogram (either by weight or by number), a very pronounced break is found at 150 km/sec, indicating that this value is close to the minimum velocity of the solar wind.

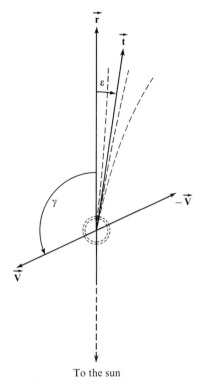

To the sun

FIGURE 4.3

The comet geometry projected into the
plane of the comet's orbit. The aberra-
tion angle $\epsilon$ and the angle $\gamma$ are shown.
(M. J. S. Belton and J. C. Brandt from
the *Astrophysical Journal Supplement*,
No. 117, Copyright 1966 by The
University of Chicago.)

A minimum velocity of the solar wind with a value $\sim 100$ km/sec was pre-
dicted by Axford, Dessler, and Gottlieb, as discussed in Section 3.7. The comet
observations have an intrinsic dispersion of about 3–4°, and this dispersion
produces a minimum velocity that is too low (due to the smearing of a relatively sharp boundary). When this effect is included, a lower bound to the
solar wind velocity of about $225 \pm 50$ km/sec is obtained. This value is
consistent with spacecraft measurements.

Large values of $w_r$ tend to produce small values of $\varepsilon$, which could be
strongly influenced by an intrinsic dispersion in velocities and errors in
measurement or reduction. The aberration angle $\varepsilon$ can be negative for some
observations. An acceptable averaging scheme must correctly include all
observations, and averaging the quantity $(1/w_r)$ satisfies this. Note that the

behavior of $(1/w_r)$ is regular and continous near $\varepsilon = 0$ while $w_r$ is not. Direct averages of $(1/w_r)$ have been computed, but the same result can be obtained in terms of averages of tan $\varepsilon$, which is approximately proportional to $(1/w_r)$. It is desirable to show the influence of $w_\phi$ on $w_r$, and, hence, we write equation (4.2) separately for direct and retrograde comets and then average them. The solar wind parameters ($w_r$ and $w_\phi$) are taken as constant, and $V \cos \gamma$ is neglected since it is much smaller than $w_r$. Add the two resultant equations to obtain

$$\langle 1/w_r \rangle^{-1} = \frac{\langle V \sin \gamma \rangle_R + \langle V \sin \gamma \rangle_D}{\langle \tan \varepsilon \rangle_R + \langle \tan \varepsilon \rangle_D}$$
$$- \langle w_\phi \rangle \frac{\langle |\cos i'| \rangle_D - \langle |\cos i'| \rangle_R}{\langle \tan \varepsilon \rangle_R + \langle \tan \varepsilon \rangle_D} \tag{4.3}$$

The azimuthal velocity enters into the determination only if the average inclinations differ between the direct and retrograde comets in the sample.

Similarly, we can subtract the two equations to obtain

$$\langle w_\phi \rangle = \frac{\langle \tan \varepsilon \rangle_R - \langle \tan \varepsilon \rangle_D}{\langle |\cos i'| \rangle_R + \langle |\cos i'| \rangle_D} \left\langle \frac{1}{w_r} \right\rangle^{-1}$$
$$- \frac{\langle V \sin \gamma \rangle_R - \langle V \sin \gamma \rangle_D}{\langle |\cos i'| \rangle_R + \langle |\cos i'| \rangle_D} \tag{4.4}$$

The radial velocity $w_r$ enters as a multiplicative factor in the first and dominant term; the second term arises because of differences in $\langle V \sin \gamma \rangle$ between direct and retrograde comets in the sample.

The average quantities needed to evaluate $w_r$ and $w_\phi$ from equations (4.3) and (4.4) are readily obtained and are tabulated for approximately 600 usable Type I comet observations in Table 4.1. Note that these values are obtained

TABLE 4.1
*Properties of Ionic Comet Tails*

| Quantity | R (retrograde) | D (direct) |
|---|---|---|
| $\langle \tan \varepsilon \rangle$ | $0.086 \pm 0.003$ | $0.062 \pm 0.002$ |
| $\langle |\cos i'| \rangle$ | 0.72 | 0.90 |
| $\langle V \sin \gamma \rangle$ | 31.7 km/sec | 34.8 km/sec |

with a specific form of the weighting function, which produces a mean $Kp$ associated with the comet observations close to the time average for $Kp$. The quantities given in Table 4.1 yield $\langle 1/w_r \rangle^{-1} = 450 \pm 11$ km/sec, and

$\langle w_\phi \rangle = 8.4 \pm 1.3$ km/sec. The radial velocity is in good agreement with space probe results. The technique described produced the first measurement of $w_\phi$, but convincing independent verification of this value is not yet available. The relatively large value implies a significant loss of solar angular momentum in the solar wind over geological time scales (see Sections 7.2 and 7.3).

The variation of solar wind velocities with geomagnetic activity (e.g., $Kp$, the planetary index) can, in principle, be investigated from cometary results. However, there is a severe practical difficulty because normally the earth and the comet are very far apart in space. When the comparisons are made, one finds that both $w_r$ and $w_\phi$ appear to increase with Kp. For example, during geomagnetically quiet times $w_\phi$ is about $4 \pm 2$ km/sec; for disturbed times the value is much larger. The variation of $w_r$ is qualitatively similar to the relationship found from the Mariner 2 data (Section 5.6).

The dispersion in the solar wind can be studied from the comet data in two ways. First, the dispersion in aberration angles of $4°.7$ r.m.s. can be analyzed. Besides any intrinsic dispersion, there are also effects as a result of errors of measurements and variations in $w_r$, $V \sin \gamma$, etc., as shown in equation (4.2). When these " spurious " effects are estimated, an intrinsic dispersion of about $3°.5$ r.m.s. remains. Second, the position angles of comet tails observed when the earth is close to the comet's orbital plane can be analyzed directly to give the dispersion and a value between 3 and $4°$ r.m.s. The second method gives the dispersion perpendicular to the comet's orbital plane while the first method gives the dispersion in the plane. As was expected these determinations indicate a roughly isotropic dispersion. The dispersions discussed here are compared with the dispersion observed by the Vela satellites in Figure 4.4. The agreement is quite good and indicates that some confidence can be placed in the cometary results. If the dispersions are caused by a peculiar velocity isotropic in the reference frame moving at the bulk speed, a velocity of about 50 km/sec is indicated.

An interesting result from the studies of comets observed while the earth was nearly in their orbital plane concerns the assumption that the comet tail lies in the orbital plane. The r.m.s. dispersion, as noted above, is not large, and the systematic departure of the tails from the orbital plane is about $1°$. It is unlikely that such departures from the orbital plane could adversely influence the results because of the weighting procedure which discriminates against unfavorable geometries; obviously very bad geometries [such as Comet Daniel (1907d)] are not analyzed for aberration angle.

Pflug has applied the type of analysis described above to the problem of the variation of the solar wind velocity with heliocentric latitude. The velocities are averaged by solar latitude of observation, and a variation is found with a minimum velocity in the latitude of the sunspot zone; such a result could be understood if the magnetic fields in the zone actually inhibited the

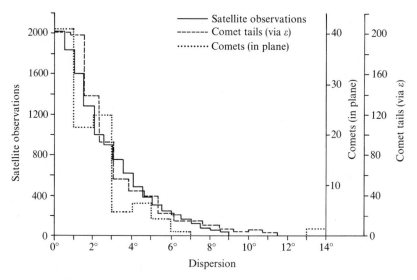

FIGURE 4.4

The dispersion in tail orientations determined both in the orbital plane (from aberration angles) and perpendicular to the orbital plane (from observations made when the earth was in the orbital plane) is compared with the dispersion in solar wind directions obtained from the Vela probes.

emission of the solar wind plasma. There are, however, two difficulties. (1) The statistics of comet observations are biased and treacherous. Such a variation could be explained by comets at different latitudes simply being observed at times of different solar activity. (2) The analysis assumes that plasma observed (via a comet observation) at a given latitude left the sun at the same latitude; this may not be true, and there is even evidence that it is not (Section 5.6). Thus, the latitude variation of $w_r$ must be treated with some caution (see the discussion of the results from interplanetary scintillations presented in Section 4.3).

## 4.2 Radar Observations

Radar reflections in the frequency range 25–50 Mc/s have been obtained directly from the solar corona by a group led by James, and they provide a valuable and independent source of data concerning the inner parts of the solar wind.

Typical observations were taken at a frequency of about 38 Mc/sec, and the radar returns showed a large spread and a Doppler shift corresponding to motion of the target toward the earth; the latter observation is a direct verification of the expansion of the corona near the sun. The signal is reflected where the local electron density achieves the value corresponding to the critical frequency or an index of refraction, $n = 0$ [equation (2.24)]. Because of the irregularity in the coronal electron distribution, the heliocentric distance of reflection is not well determined, but it is probably in the range 1.5–2.0 $R_\odot$. The expansion velocity is about 16 km/sec, a value which is compatible with our knowledge of the solar wind from other sources. The mean spread in return frequencies corresponds to a mass motion of 35 km/sec in the radial direction. Such mass motions could influence coronal temperatures determination via Doppler widths (on the limb) by about $0.25 \times 10^6$ °K.

Reflections are also observed from "irregularities" moving at speeds $\sim 100$–200 km/sec. These cannot be bulk motions (because the equation of continuity would give huge fluxes,) and the irregularities are probably the shocks responsible for the heating of the corona. If so, the results are particularly interesting because these irregularities appear to slow down or die out near $r = 1.6\, R_\odot$, a value in agreement with the value quoted in Section 2.3 from a study of the coronal density gradient.

Radar reflections can be obtained easily from the moon, and the plasma between the earth and the moon clearly influences the returns. However, the influence of the ionosphere is large and variable, and the information concerning the cis-lunar plasma is completely masked. However, the technique should be applicable to the studies of other planets where the column electron density is increased by orders of magnitude and the influence of the ionosphere is negligible.

## 4.3   Radio Observations

Analysis of radio radiation from various astronomical sources can provide valuable information concerning the solar system plasma through which the radio waves must propagate. Celestial sources such as the Crab Nebula and Jupiter have been used.

The variation in apparent size of the Crab Nebula as its position moves through the solar corona provides information about the scale and orientation of irregularities in the electron density. A point source viewed through an irregular transparent medium increases in size because of the cumulative effect of many small deviations as irregularities are encountered in the medium. The apparent diameter of such a point source can be calculated from geometrical optics and is approximately

$$2\phi \approx 3 \times 10^{-4} \frac{\delta N_e n^{1/2}}{f^2} \tag{4.5}$$

Here $\phi$ is in radians, $\delta N_e$ is the r.m.s. electron density variation either above or below the surrounding medium in electrons/cm³, $f$ is the frequency in

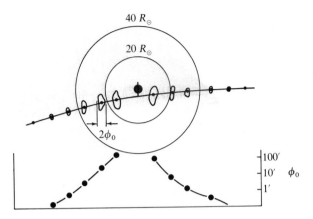

FIGURE 4.5
Schematic of the appearance of the Crab Nebula as it
passes through the solar corona. The bottom scale is
appropriate to observations at 38 Mc/s.

Mc/s, and $n$ is the number of irregularities or clouds along the line of sight.
The magnitude of the effect can be estimated with some typical parameters.
Consider a ray passing within about 20 $R_\odot$ of the sun with about $10^2$ clouds
along the line of sight; take $\delta N_e \approx N_e$ at the point of closest approach or
$\delta N_e \approx 10^3/\text{cm}^3$. These values and $f = 100$ Mc/s yield a typical dispersion
($2\phi$) of about 1 min of arc; this value increases greatly as the line of sight
moves closer to the sun and as the frequency is lowered. The observed value
of $\phi \approx 10'$ at 20 $R_\odot$ and 38 Mc/sec is reproduced from equation (4.5) when an
increase ($\times 3$) is included to allow for radial variation of parameters. The
schematic appearance of the Crab Nebula is shown in Figure. 4.5. The
scattering angle has been observed over the frequency range 26 to 178 Mc/s,
and it varies as $f^{-2}$; this indicates that the multiple scattering picture [on
which equation (4.5) is based] is probably realized in practice. Scattering has
been observed out to approximately 0.5 a.u.

Comparison of the details of the scattering observations with calculations
based on different coronal models has been carried out. The scale of the
irregularities cannot be accurately determined, but the limits are between 10
and about 5,000 km. Since radio scattering is always present, the irregularities
are always present. The logical mechanism to produce ever-present irregularities
with this scale is the magnetic field; recall that the magnetic field is thought
to be responsible for most of the small-scale fine structure in the corona
(Section 2.4).

If the magnetic field is the agent responsible for the irregularities in the
density, we might expect (on the basis of our coronal experience) a filamentary

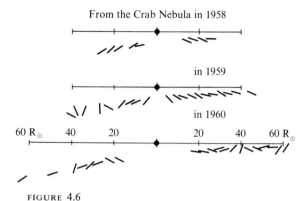

FIGURE 4.6
The inferred orientations of the coronal-interplanetary
filaments that are responsible for radio scattering.

structure elongated in the radial direction. The maximum scattering would
occur in the direction perpendicular to the axis of the filament. For the corona,
the observations indicate that the scattered radiation is distributed in an
ellipse of axial ratio 2:1 where the major axis is usually perpendicular to the
radius vector. Thus, the scattering is apparently due to the effects of magnetically confined, radially aligned irregularities. Sample orientation results
are shown in Figure 4.6. Note that the results refer to an integrated path
through the corona and that the observations are difficult to make. In order
to remove ambiguity, three triangularly oriented baselines are required.

The amount of scattering is also known to be greatly reduced in the polar
regions near sunspot minimum. Such results can be displayed by plotting
contours of equal mean scattering angle $\phi_0$. Such contours can be used to
determine the Ludendorff index [equation (2.7)] of flattening, and $\varepsilon \approx 0.4$ is
found for $r = 50 \, R_\odot$. Observations have been carried out over enough years
to determine a solar cycle variation. A general increase of a factor of two in
mean scattering angle from solar minimum to solar maximum is found within
$50 \, R_\odot$. The radial variation of $\phi_0$ is found to be $(r/R_\odot)^{-\beta}$ where $\beta$ lies between
1.3 and 2.3. Attempts to relate this variation to properties of the solar wind
flow have not been very convincing.

In addition to the relatively gross phenomena described above, there is
also the phenomenon of interplanetary scintillation consisting of fluctuations
in the flux of small diameter radio sources (such as 3C 48) with a frequency of
0.1–10 cycles/sec. Such fluctuations are attributed to the presence of small-
scale irregularities in the interplanetary electron density, moving at or near
the solar wind speed. The physical origin of such fluctuations can be seen in
Figure 4.7. Refraction caused by density fluctuations produces changes in
intensity as a function of position. These changes are carried past a stationary

114

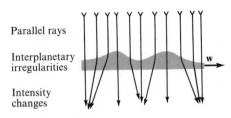

Parallel rays

Interplanetary irregularities

Intensity changes

w

FIGURE 4.7

The physical origin of interplanetary scintillations.

observer by the solar wind to produce fluctuations with time. This can be viewed as the passage of a diffraction pattern across the surface of the earth, much as moving ripples on the surface of a swimming pool produce a moving light pattern on the pool bottom. This assumption can be checked by observations at different locations. A good correlation exists for stations separated by 50 km, but the correlation vanishes when the stations are separated by a few hundred km or more. Apparently, the fluctuations persist long enough to be convected $\approx 50$ km, or a time of at least $(50 \text{ km/w}) \approx 0.1$ sec. They apparently do not last for times (say) $\sim 1$ sec or more. Thus, the irregularities are not permanent features, and wave phenomena as discussed in Section 3.9 may be the answer.

The interplanetary scintillations must be disentangled from the ionospheric scintillations caused by electron density irregularities in the ionosphere; the scale of the ionospheric irregularities is $\sim 5$ km. The correlation of the scintillations with respect to solar position (assuming a maximum contribution when a ray passes closest to the sun) and the scintillation rate established as interplanetary the 0.1–10 cycles/sec phenomenon.

With the aid of certain reasonable assumptions, the detailed observations can be analyzed and a scale of a few proton cyclotron radii ($\sim 10^2$ km) found; the amplitude of the fluctuation is $\approx 2$ percent of the mean electron density. Disturbances apparently caused by solar flares have been reported. Observations made with three receivers enable a determination of the projected speed and direction of the solar wind. Dennison and Hewish studied the region from 0.36–0.80 a.u. and found average speeds in the range 200–600 km/sec, as shown in Figure 4.8. It is believed that the variation in Figure 4.8 could result from a variation of the speed of the solar wind with the solar latitude as possibly indicated by the comet tail observations (see discussion in Section 4.1). Note, however, that the details of the latitudinal variation of $w$ found from comet tails and interplanetary scintillations are quite different. The $w$ from comet tail studies shows a minimum in the sunspot zones while the $w$ from scintillation studies increases monotonically up to latitudes of 60° (the limit of the observations). Observations of interplanetary scintillations in

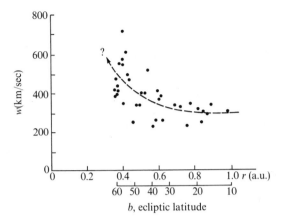

FIGURE 4.8
The solar wind velocity $w$ as a function of the heliocentric coordinates $r$ and $b$ of the point closest to the
sun along the line of sight. (After P. A. Dennison
and A. Hewish.)

radio radiation from Jupiter are also consistent with a solar wind speed of a
few hundred km/sec.

The observations of the type presented by Dennison and Hewish are of
great value because they refer to regions of the solar wind that have not been
explored by deep space probes. Comets, as discussed in Section 4.1, probe
essentially the entire interplanetary region, but they come at irregular and
generally unpredictable times. The radio observations can be carried out on a
regular or routine basis. A comparison of costs would certainly weigh heavily
in favor of the study of interplanetary scintillations as a powerful method for
probing the velocity field of the solar wind.

## 4.4 Geomagnetic Observations and Cosmic Rays

The earth's magnetic field is immersed in the solar wind, and, with a proper
theory, properties of the solar wind could be inferred from variations in the
earth's magnetic field at the surface. The magnetic field variations have been
measured at the surface of the earth for decades and provide a potentially
valuable fund of data; obvious variations such as those resulting from diurnal
effects and atmospheric tides, etc., must, however, be removed from the data.

Evidence for a correlation between certain properties of the solar wind and
geomagnetic indices such as $Kp$ will be presented and discussed in Section
5.6. A correlation of $Kp$ variations with the sector structure discussed in

Section 5.6 could be useful in providing a history of the sector boundaries (but in practice this may be difficult). The physics of the interaction between the solar wind and the geomagnetic field is not well understood. Straightforward hypotheses depending, for example, on variations in the impact pressure of the solar wind are not adequate to explain the observed variations. Recent results based on data from Mariner 4 indicate an excellent correlation between geomagnetic variability ($Kp$) and the magnitude of the fluctuations in the interplanetary magnetic field in the plane perpendicular to the earth-sun line.

One generally accepted method for inferring some solar wind properties uses the Chapman-Ferraro theory for the initial phase of geomagnetic storms. If a velocity of $10^3$ km/sec is taken from the delay times, an initial phase amplitude of $50\gamma$ is produced by an enhanced solar wind density of $10/cm^3$.

Delay times between solar and geomagnetic events can be used to infer velocities for flare events and possibly M-region streams (see Section 1.2). Some information concerning the shape of flare ejecta is contained in the geomagnetic records. Flares anywhere on the visible solar disk are capable of producing a sudden storm commencement (SSC) via the blast wave discussed in Section 3.8, but the magnitude of the main phase is strongly dependent on disk position. The fact that flares near the disk center produce the largest storms suggests a narrow tongue-like ejection. The same impression is apparent from the time delay between the SSC and the onset of the main phase because the standoff distance of the shock depends on the curvature of the cloud of driver gas (Section 3.8).

The inverse correlation between sunspot number and cosmic ray intensity (see Sections 1.2 and 6.7) as measured by monitors on the earth's surface contains information concerning the extent of the region responsible for the modulation. If the cosmic ray variation lagged behind the sunspot indices by three months, this could be interpreted as the time for the solar wind to travel from the sun to the boundary of the modulating region; if $w = 500$ km/sec, the boundary comes at $r = 30$ a.u. It was thought at first that this was the case, but it was found that the delay is greatly reduced if the coronal green line intensity is used to measure solar activity. Apparently, the delay was primarily between the photosphere and the corona; delays on the order of months follow from the known fluxes and the equation of continuity. Hence, the size of the modulation region is greatly reduced. It could be comparable with the 5 a.u. dimension suggested by Hundhausen (Section 3.10) if the dimensions of the supersonic solar wind and the modulating region are essentially the same. However, there is no firm reason to believe that the dimensions of these regions are similar. Apparently little information concerning the size of the modulating region is to be found in the type of study described above.

# Bibliographical Notes: Chapter 4

*General References*

A review of most of the methods described in this chapter is contained in:
1. Axford, W. I.: *Space Sci. Revs.*, **8**, 331 (1968).
2. Lüst, R.: in *Solar-Terrestrial Physics*, J. W. King and W. S. Newman, Eds., Academic Press, London (1967), p. 1.

*Section 4.1*

See the references for Section 1.3 and:
3. Brandt, J. C.: *Ann. Rev. Astron. Astrophys.*, **6**, 267 (1968).
4. Lüst, Rh.: *Zs. f. Ap.*, **51**, 163 (1961).
5. Biermann, L., and Rh. Lüst: in *Nature et Origine des Comètes* (13th Liège Symposium, Cointe-Sclessin, Belgium, Institut d'Astrophysique) (1966), p. 329.
6. Belton, M. J. S., and J. C. Brandt: *Ap. J. Suppl.*, **13**, 125 (No. 117) (1966).
7. Brandt, J. C.: *Ap. J.*, **147**, 201 (1967).
8. Brandt, J. C.: *A. J.*, **73**, S168 (1968).
9. Pflug, K.: *Publ. Ap. Obs. Potsdam*, No. 106 (1966).

*Section 4.2*

Discussions of radar results are found in:
10. Chisholm, J. H., and J. C. James: *Ap. J.*, **140**, 377 (1964).
11. Brandt, J. C.: *Science*, **146**, 1671 (1964).
12. James, J. C.: *Ap. J.*, **146**, 356 (1966).
13. Brandt, J. C.: *Ap. J.*, **149**, 447 (1967).

*Section 4.3*

A review of radio studies of the corona and solar wind is found in:
14. Hewish, A.: in *Solar System Radio Astronomy*, J. Aarons, Ed., Plenum Press, New York (1965), p. 255.

See also:
15. Slee, O. B.: *Planetary Space Sci.*, **14**, 255 (1966).

Interplanetary scintillations are discussed in:
16. Dennison, P. A., and A. Hewish: *Nature*, **213**, 343 (1967).
17. Douglas, J. N., and H. J. Smith: *Ap. J.*, **148**, 885 (1967).
18. Slee, O. B., and C. S. Higgins: *Aust. J. Phys.*, **21**, 341 (1968).

*Section 4.4*

See the discussion and references for Section 5.6, Reference 1, and:
19. Ballif, J. R., D. E. Jones, P. J. Coleman, L. Davis, and E. J. Smith: *J. Geophys. Res.*, **72**, 4357 (1967).

# 5

# Space Observations

Our knowledge of the properties of the solar wind and the basis of any theoretical discussion of the subject rests on direct measurements by space probes. Such *in situ* measurements have been possible since 1961. In 1962, Mariner 2 produced essentially continuous measurements of the solar wind covering a period of about five solar rotations and found that a detectable flux was always present. Thus, the existence of a continuous solar outflow or solar wind was established beyond any reasonable doubt. (The history of direct measurements was briefly summarized in Section 1.3.)

In this chapter, we will first review the principal results of direct measurements and then summarize them (Section 5.8). In addition, we will briefly discuss the physical principles of the instruments used to make space observations of the solar wind plasma.

## 5.1 Densities

Most direct measurements of densities by plasma probes refer to the ion or proton component. The combination of a flux ($N_p w$) and a velocity determination from the energy per unit charge ($E/Q$) gives the density. The measured quantity is usually in terms of $E/Q$ and, hence, is not unique.

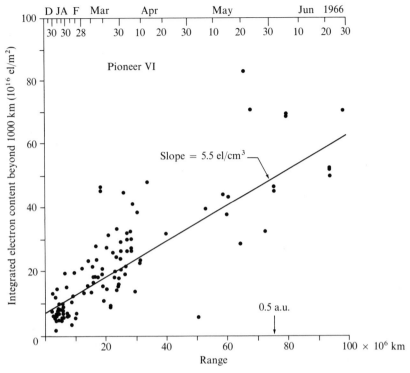

FIGURE 5.1

The column electron density as a function of range as determined from measurements made with Pioneer 6. The solid line has a slope corresponding to a mean electron density of 5.5 electrons/cm³. Note that m.k.s. units are used in this figure. (Courtesy H. T. Howard and R. L. Koehler.)

The direct measurements of the electron component of the solar wind plasma are greatly complicated by the potential on the spacecraft and by the presence of photoelectrons; also, the solar wind velocity is greater than the mean proton thermal velocity, but less than the mean electron thermal velocity. Currently these problems appear to have been overcome. For some time, observations of interplanetary electron density depended upon radio propagation effects in signals transmitted between earth and deep space probes. A determination of the electron density is possible because the group and phase velocities depend on frequency and electron density; therefore, pulses sufficiently different in frequency would show a time delay after traversing a large column density of electrons, or continuous signals would show a difference in phase. For the large paths encountered in practice, the ionospheric effects become unimportant.

Sample results from the radio propagation experiments are shown in Figure 5.1. (Note that m.k.s. units are used in this figure.) Results from

Pioneer 6 (December 1965 to May 1966) are an $\langle N_e \rangle$ of 4.3/cm$^3$ with an r.m.s. of 3.6/cm$^3$. Results from Pioneer 7 (August 1966 to March 1967) give an $\langle N_e \rangle$ of 8.7 $\pm$ 4.0/cm$^3$. These results are entirely compatible with space-craft determinations and the assumption of $N_e = N_p$. These independent determinations of the electron density are fundamental, particularly because a large section of space is probed to determine a meaningful average. Even density enhancements of a factor of $\approx 6$ have been measured by this technique.

The technical problems of direct electron observations have been overcome, and results were published in 1968. Some sample data by Montgomery, Bame, and Hundhausen are shown in Figure 5.2; these data are from the Vela 4B electrostatic analyzer. The proton and electron densities and bulk velocities are equal to within the errors of measurement. Note that $N_e$ should be slightly larger (by $\sim 10$ percent) than $N_p$ because of the presence of helium in the solar wind.

Observed electron or proton densities range from about 0.4 to 80/cm$^3$ with an average near 5/cm$^3$. These values are based on a variety of independent methods of measurement.

## 5.2   Velocities

The bulk velocity of the solar wind plasma is inferred from the distribution of energy per unit charge ($E/Q$) as measured by the plasma probe. A sample spectra from the Vela 3A satellite is shown in Figure 5.3. The peak near 1 kev is due to protons, and the peak near 2 kev is due to alpha particles moving at the same bulk speed (and thus having twice the energy per unit charge). Most detectors measure the number of particles in a range of energy per unit charge. If the size of the energy interval is sufficiently narrow, the peak of the distribution is well determined. Clearly, the velocity corresponding to the energy of a proton at the peak of the distribution must be very close to the bulk velocity. [Observations of mass per unit charge ($M/Q$) have been carried out directly by Ogilvie, Burlaga, and Wilkerson.]

Bulk velocities in the range 200–900 km/sec have been measured. The over-all average is in the range 400–500 km/sec while the average for "quiet" or undisturbed periods is probably between 300 and 350 km/sec. Sample data for electron bulk velocities are shown in Figure 5.2.

A histogram showing the distribution of solar wind velocities over a period of one year as obtained from satellites Vela 2A and 2B is given in Figure 5.4. The observations were taken during times of relatively quiet activity, and the mean velocity of 420 km/sec is marked. The Mariner 2 observations were taken during a period of higher activity, and the mean velocity obtained was 504 km/sec. The behavior of the average velocity of the solar wind with the solar cycle depends on the frequency of high velocity streams. The quiet-time base level of velocity is $\approx 325$ km/sec, and $w$ rises above this in high

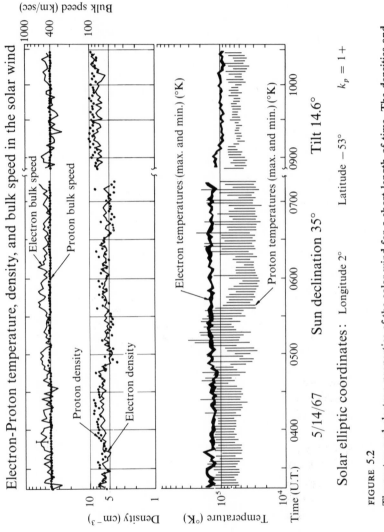

Electron-Proton temperature, density, and bulk speed in the solar wind

FIGURE 5.2

The proton and electron properties of the solar wind for a short length of time. The densities and velocities of protons and electrons are equal to within the errors of measurement, but the temperature differences are significant. (Courtesy M. D. Montgomery, S. J. Bame, and A. J. Hundhausen.)

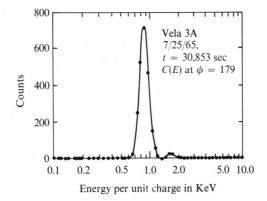

FIGURE 5.3

A typical energy spectrum showing the major peak, which is due to protons, and a secondary peak, which is due to alpha particles. (Courtesy A. J. Hundhausen, J. R. Asbridge, S. J. Bame, and I. B. Strong.)

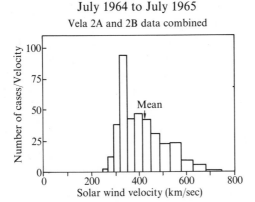

FIGURE 5.4

The frequency distribution of solar wind velocities observed on Vela 2A and 2B. (Courtesy I. B. Strong, J. R. Asbridge, S. J. Bame, and A. J. Hundhausen.)

velocity streams, which were common during the time of the Mariner 2 observations and rare during the time of the Vela 2A and 2B observations.

The bulk velocity is usually taken to be very nearly the radial ($r$) component of the solar wind velocity. Possible components in the $\phi$ (azimuthal) and $\theta$ (polar) directions are also of great interest. These can be determined by observing the distribution of counts at a fixed $E/Q$ as a function of direction in a spinning spacecraft (see Figure 6.3). Such measurements show a "random" fluctuation in direction of the solar wind with an average value $\sim 5°$ (see Figure 4.4). Departures from the average direction can go as high as $\approx 10°$–$15°$. In addition, the systematic azimuthal component of velocity ($w_\phi$) comparable in value with the determination from the comet studies (Section 4.1) appears to have been obtained. The significance of any systematic $w_\phi$ is discussed in Section 7.3.

## 5.3   The Velocity Distribution and Temperature

The temperature of the solar wind is defined in terms of the velocity dispersions in the frame of reference moving at the bulk speed of the solar wind. The temperature in a given direction ($\Phi$) can be defined in terms of the mean square velocity ($v^2$) in that direction, viz.,

$$T(\Phi) = \frac{m\langle v^2 \rangle}{k} \tag{5.1}$$

Thus, the temperature is essentially given by the width of the peak in the distribution of $E/Q$ in Figure 5.3. Note, however, that the finite resolving power of the instrument also contributes to the width. The average proton temperature quoted is $\approx 10^5$ °K; extreme values are a maximum $\sim 10^6$ °K and a minimum of about $5 \times 10^3$ °K.

Sample electron temperatures are shown in Figure 5.2. The electron temperatures average about a factor of three to four times the proton temperature (compare with the theoretical values in Section 3.5). The coupling between protons and electrons is apparently larger than was anticipated, but not large enough to ensure temperature equality. The first indication of the temperature of the solar wind electrons appears to have come from Burlaga's study of the velocity discontinuities in the flow. The total pressure must be constant across such contact surfaces, and hence $(B^2/8\pi) + N_e k(T_p + T_e) = $ const. A plot of $B^2$ versus $N_e$ has a slope proportional to $(T_p + T_e)$; since $T_p$ was measured directly, the electron temperature $T_e$ was determined.

A great deal of information can be displayed in the form of contour maps in a plane of velocity space. An example is shown in Figure 5.5, which is based on data from the satellite Vela 3B. The $v_1 v_2$ plane is defined by the velocity in the radial direction ($v_1$) and the velocity ($v_2$) perpendicular to both the radial direction and the spin axis of the satellite. The contours are drawn

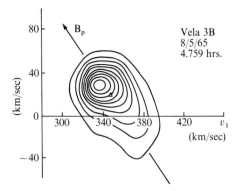

FIGURE 5.5

A contour map in the $v_1v_2$ plane obtained from
Vela 3B; the small triangle denotes the bulk
velocity. The direction of the magnetic field
(obtained from IMP-3) projected into the
$v_1v_2$ plane ($B_p$) is shown for comparison.
(Courtesy A. J. Hundhausen, S. J. Bame, and
N. F. Ness.)

in steps of one-tenth the maximum value. The bulk velocity is marked by a
triangle. If the velocity distribution were isotropic, the contours would be
concentric circles centered on the bulk velocity. An appreciable anisotropy
is readily evident and clearly aligned with the direction of the magnetic field
which has been projected into the $v_1v_2$ plane. Comparison of data concerning
magnetic field direction from IMP-3 and anisotropy direction from Vela 3
indicates that the two directions are generally instantaneously aligned.

The anisotropy can be expressed in terms of the temperature as a function
of direction. This is shown in Figure 5.6. The ratio $T_{\parallel}/T_{\perp}$ is about 3.4 in the
example shown. Parallel and perpendicular refer to the direction of the magnetic field. The average temperature anisotropy shows $T_{\parallel}/T_{\perp} \approx 2$. The temperature anisotropy can be explained by the first adiabatic invariant, $\mu$,
where

$$\mu = \frac{\frac{1}{2}mv_{\perp}^2}{B} \tag{5.2}$$

Here $m$ is the mass of the particle, $v_{\perp}$ is the velocity perpendicular to the
magnetic field, and $B$ is the field. Even if the velocity distribution is isotropic
near the sun, the component perpendicular to the field must decrease as $B$
decreases with distance from the sun. This tendency is offset by coupling
mechanisms (waves, instabilities, etc.) in the plasma.

Figures 5.5 and 5.6 show one additional characteristic, namely, that the
velocity dispersion or temperature of particles traveling away from the sun is

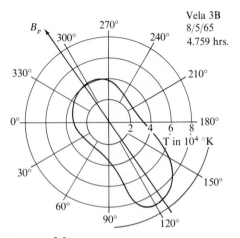

FIGURE 5.6

The temperature $T(\Phi)$ as defined by equation (5.1) is shown in the $v_1 v_2$ plane for the distribution measured in Figure 5.5; the projected field $B_p$ is also shown. (Courtesy A. J. Hundhausen, S. J. Bame, and N. F. Ness.)

larger than the velocity dispersion or temperature of particles traveling toward the sun. Note that the magnetic field is directed toward the sun in Figures 5.5 and 5.6. Skewed distributions are generally the case regardless of the field direction, and the phenomenon can be interpreted as evidence of heat conduction by the ions away from the sun. The skewed distributions that are typically observed imply an energy transport of $\sim 10^{-5}$ ergs/cm$^2$-sec. This is several orders of magnitude smaller than the conductive energy transport by electrons, which is $\sim 10^{-2}$ ergs/cm$^2$-sec and which has now been observed.

The "temperature" of the alpha particles in the solar wind has been measured and is about four times the proton temperature. This means that the peculiar velocities of protons and alpha particles are equal and suggests a "turbulent" regime rather than the usual thermal processes. This is indicative of a nonequilibrium situation, and the definition of temperature in equation (5.1) should be kept clearly in mind.

Finally, electron temperatures can be inferred (at least in principle) by studies of Type III bursts at low frequencies $\approx 1$ Mc/s from satellites above the ionosphere; the method is the same as that described in Section 2.4. The lower frequencies permit probing of the solar wind at heliocentric distances of some 10 to 55 $R_\odot$. Unfortunately, the decay times lead to very low temperatures. The situation appears to worsen with increasing heliocentric distance, and some facet of the theory or analysis becomes progressively less valid.

## 5.4   Composition

While we have often spoken of the solar wind as a proton-electron gas, heavier ions are expected in the plasma; one of these is alpha particles, which are the most abundant. Alpha particles have been detected since the flight of Mariner 2 in 1962; Figure 5.7 shows the distribution of measured values of $N(\alpha)/N(p)$ obtained from Vela satellites in 1965. The average value (based on several determinations) is $N(\alpha)/N(p) = 0.046$, but note the very large variation.

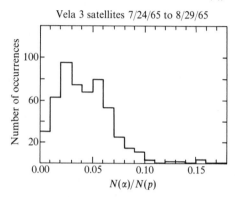

Distribution of measured ratios $N(\alpha)/N(p)$

FIGURE 5.7

The distribution of $N(\alpha)/N(p)$ values obtained from Vela 3 satellites. (Courtesy I. B. Strong, J. R. Asbridge, S. J. Bame, and A. J. Hundhausen.)

This variation implies a different ease of escape for protons and alpha particles; the maximum value of $N(\alpha)/N(p) \approx 0.2$ could be the coronal value observed at the earth for a case of bulk ejection. The entire corona is replenished within a time on the order of weeks and can respond to imbalances in the incoming (from the photosphere) and outgoing (into the solar wind) chemical composition rather quickly. On the average, the relative abundance of a heavy element varies until the coronal chemical composition reaches a steady state. If it is more difficult for helium and iron to escape in the solar wind, we would expect their abundances to increase in the corona. This appears to be the case for iron, as discussed in Section 2.4. Also, the average of abundances by flux in the solar wind should represent the photospheric or input values to the corona. The solar wind value for helium of $N(\alpha)/N(p) = 0.046$ is not far from an indirect photospheric value of 0.06 obtained from a method based on solar cosmic ray abundances (Section 2.4). Ogilvie, Burlaga,

128

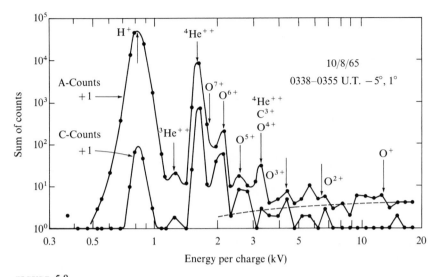

FIGURE 5.8

The $E/Q$ distribution of positive ions showing some heavy elements, notably oxygen; the dashed curve indicates the level of background counts. (Courtesy S. J. Bame, A. J. Hundhausen, J. R. Asbridge, and I. B. Strong.)

and Wilkerson find a marked increase in the helium abundance after the main phase of some (presumably flare-induced) geomagnetic storms.

An instrument with sufficient sensitivity and range could observe heavier elements in the solar wind. Such an instrument has been flown on the Vela series of satellites, and a sample of the distribution of $E/Q$ for positive ions is shown in Figure 5.8. This spectrum and others like it which resolve the various ion species of $^{16}O$ are obtained when $T_p \approx 10^4$ °K, values which are unusually low and near the minimum for the solar wind. In addition to the usual $H^+$ and $^4He^{++}$, the Vela measurements show the presence of $^3He^{++}$, $^4He^+$, $O^{+5}$, $O^{+6}$, and $O^{+7}$ as well as other (unidentified) heavy ions; the peaks corresponding to these ions are shown in Figure 5.8. The dashed curve gives the average level of background counts. A summary of the heavy ion abundance determinations is given in Table 5.1.

The relative abundances of $O^{+6}$ to $O^{+7}$ can be used to infer the coronal temperature because, as has been known for some years, the ionization state of the solar wind is established deep within the corona. The variation of the fraction of oxygen in various stages of ionization is shown in Figure 5.9 which was calculated by Hundhausen, Gilbert, and Bame using the solar wind model published by Whang and Chang. Coronal temperatures of $2.0 \times 10^6$ °K or a little less are obtained. Note, however, that these determinations refer to times at the earth when the solar wind plasma is rather cool ($\sim 10^4$ °K) and,

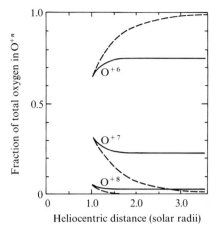

FIGURE 5.9

The distribution of ionization states of oxygen in the solar wind. The solid lines are results based on the solar wind model of Whang and Chang while the dashed line shows results from a static model (but with the temperature distribution from the Whang and Chang model). (Courtesy A. J. Hundhausen, H. E. Gilbert, and S. J. Bame.)

hence, may not be entirely typical. The stage of ionization of the solar wind appears to be a unique property in that it is almost unaffected by the tremendous expansion of the coronal material.

The detection of $^3He^{++}$ ions in the solar wind appears to be the first observational evidence for $^3He$ in the sun.

The observation of $^4He^+$ in the solar wind with an abundance $\sim 10^{-3}$ $^4He^{++}$ could be an anamoly because such a relatively high abundance of singly ionized helium would not be expected in the corona. Three possible resolutions of this anomaly are under consideration: (1) the solar wind does not

TABLE 5.1
*Estimated Ion Species Ratios*

| Date | $^3He/^4He$ | $^4He^+/^4He^{++}$ | He/O | $O^{+7}$ : | $O^{+6}$ : | $O^{+5}$ : |
|---|---|---|---|---|---|---|
| March 22, 1966 | $\leq 4 \times 10^{-4}$ | $<10^{-3}$ | 80 | 2 | 1 | 0.1 |
| March 23, 1966 | $\leq 2 \times 10^{-4}$ | $\leq 10^{-3}$ | 50 | 3 | 1 | $\leq 0.1$ |
| Oct. 8, 1965(1) | $1.3 \times 10^{-3}$ | $\approx 3 \times 10^{-3}$ | 25 | $\leq 0.2$ | 1 | 0.1 |
| Oct. 8, 1965(2) | $1.3 \times 10^{-3}$ | $\approx 3 \times 10^{-3}$ | 66 | 0.3 | 1 | 0.1 |

SOURCE: After Bame, Hundhausen, Asbridge, and Strong (1968).

originate entirely in the corona; (2) the $He^+$ originates by charge-exchange of $He^{++}$ with neutral hydrogen in the interplanetary medium; or (3) the $He^+$ originates by charge-exchange of $He^{++}$ with terrestrial neutral hydrogen in an extended cloud around the earth. This latter possibility must be considered because the Vela probes are in a nearly circular orbit with a radius of about 17 $R_E$. The first possibility appears unlikely because the ratio $O^{+7}/O^{+6}$ is entirely consistent with the solar wind originating in the corona. It is difficult to favor either the second or third hypothesis. The second would imply far more neutral hydrogen (presumably of interstellar origin) in the interplanetary medium than expected from our discussion in Section 3.10. The third would appear to require a more extended and denser terrestrial hydrogen cloud than is currently considered plausible. Direct observation of neutral hydrogen atoms in the solar wind or the interplanetary medium should resolve this problem.

## 5.5   The Magnetic Field

The interplanetary magnetic field is an integral part of the solar wind and is the photospheric field extended outward by the expansion of the plasma. From the theory presented in Sections 3.1 and 3.7, we would expect an average field near earth of about $5\gamma$; the solar rotation leads to the development of the characteristic Archimedean spiral with an angle of about $45°$. Available observations confirm this general picture.

A histogram (based on observations covering three solar rotations; see Figure 5.10) shows the distribution of field magnitude that was obtained by Ness, Scearce, and Cantarano from IMP-1 data. The median value is $5.5\gamma$

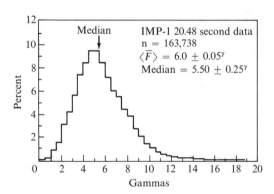

FIGURE 5.10

The distribution of interplanetary magnetic field magnitude as observed by IMP-1 from 1963 to 1964. (Courtesy N. F. Ness.)

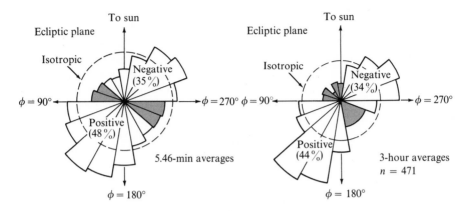

FIGURE 5.11
The distribution of interplanetary field directions observed by IMP-1; note the change when a longer averaging period is used. (Courtesy N. F. Ness, C. S. Scearce, J. B. Seek, and J. M. Wilcox.)

($1\gamma = 10^{-5}$ gauss). Extreme values of the field magnitude are a low of about $0.25\gamma$ and a high near $40\gamma$. Vector magnetic fields can currently be measured to within $\pm 0.25\gamma$; such an accuracy is dependent on the construction of non-magnetic spacecraft.

The directional properties of the magnetic field are determined by plotting histograms of field direction as projected into the plane of the ecliptic (see Figure 5.11); the $7°$ difference between the plane of the solar equator and the plane of the ecliptic is unimportant for this discussion. The clear tendency of the field to lie along angles in a solar-ecliptic coordinate system of $135°$ or $315°$ (corresponding to a spiral angle of $45°$) is shown, particularly when the results are averaged over a longer time period. Hence, a gross overall Archimedean spiral geometry is indicated, but with a substantial amount of local irregularity.

Measurements made on board Mariner 2, Pioneer 5, and IMP-1 have indicated a southward component of field perpendicular to the plane of the ecliptic amounting to about $1\gamma$. If such a permanent component persists across a sector boundary (see Section 5.6) where the field changes direction (say) from toward to away from the sun, the angle that the field lines make with the ecliptic plane must change sign; Dessler has pointed out the disagreeable nature of this possibility. Davis has warned of the severe consequences of such a component for the solar magnetic field. If the magnetic flux is frozen-in, the time rate of change of a net magnetic flux through the plane of the ecliptic is

$$\frac{\partial \Phi}{\partial t} = 2\pi r w B_{\perp} \qquad (5.3)$$

For $B_\perp = 10^{-6}$ gauss $(0.1\gamma)$, $r = 1.5 \times 10^{13}$ cm, and $w = 4 \times 10^7$ cm/sec, we obtain a flux loss of $4 \times 10^{15}$ maxwell/sec. Over a year's time $(3 \times 10^7$ sec), this becomes $\approx 10^{23}$ maxwells or roughly the magnetic flux in one hundred average sunspots. Such a loss would be difficult to explain and is highly unlikely. An instrumental origin for the perpendicular component of the interplanetary magnetic field seems likely. Measurements carried out on Pioneer 6 (from December 1965 to September 1966) indicate a zero net perpendicular field.

## 5.6    Variations

Radial variations of the density, flux, and bulk kinetic energy are of interest in themselves as well as providing checks on our theoretical understanding. Neugebauer and Snyder have compared these quantities observed on Mariner 2 with the expected inverse square variation (Figure 5.12), and the theoretical expectation is verified. The radial variation of the total magnetic field was studied on Pioneer 6 for 0.81 a.u. $< r < 1.0$ a.u. by Burlaga and Ness and on Mariner 4 for 1.0 a.u. $< r < 1.5$ a.u. by Coleman, Smith, Davis, and Jones. The field strength for the latter experiment could be approximated by $B = 4.13 \, (r/r_0)^{-1.25}$ where $r_0 = 1.5$ a.u. and the field is in $\gamma$; an ideal spiral field in a quiet solar wind at 350 km/sec would have $B \propto (r/r_0)^{-1.29}$.

Variations in the density and solar wind velocity during the flight of Mariner 2 have been presented by Neugebauer and Snyder, and these are shown in Figure 5.13. These data clearly show that high velocity is correlated with low density, and vice versa. Thus, the flux $wN_e$ tends to be somewhat constant. Vela measurements show that high proton temperatures are observed when the bulk velocity is high and that low temperatures are observed when the velocity is low. Eventually, acceptable models of the solar wind will have to explain these gross features. Variations of $w$ with geomagnetic index $Kp$ are also known, but we defer this discussion until consideration of the "sector structure" (below).

A great deal of information is available about the variations in the magnetic field. Ness has pointed out that we have

$$\left[\frac{dB_I}{dt}\right]_{\text{observed}} = \frac{\partial B_I}{\partial t} + (\mathbf{w} \cdot \mathbf{\nabla})B_I \qquad (5.4)$$

where $B_I$ is the interplanetary field. Since the solar wind speed is some five to ten times any magnetohydrodynamic wave speed, the second term in equation (5.4) dominates. Thus, the major observed change is due to convection of different regions past the observer. With this fundamental limitation in mind, consider the interpretation of variations in (say) the magnetic field

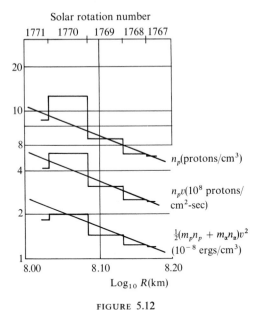

FIGURE 5.12

The radial variation of density, momentum, and
energy averaged over a solar rotation obtained
from Mariner 2. (Courtesy M. Neugebauer and
C. W. Snyder, and the *Journal of Geophysical
Research*.)

observed as a function of time. Should these variations be interpreted as
waves or discontinuities or whatever? Coleman has searched for the "sig-
natures" of different wave forms and has reported evidence for Alfvén and/or
fast mode waves. Sari and Ness have proposed a rather different interpretation. Analyses of time variations of the interplanetary magnetic field are often
carried out in terms of "power spectra" which are the square of the Fourier
transform of the observed variation. Sari and Ness note that individual discontinuities (and a random distribution of discontinuities) have a power
spectrum proportional to $(frequency)^{-2}$ and that such a dependence is evident
in data obtained from Pioneer 6 for the appropriate frequency range (2.8
$\times 10^{-4}$ to $1.6 \times 10^{-2}$ c/s) and well above the noise level. This behavior is
attributed to directional discontinuities in the microstructure (see below for
definition), and such discontinuities alone are sufficient to explain the power
spectra. At lower frequencies, the interplanetary macrostructure shows up
in the power spectra. The overall situation is not yet entirely settled. A sample
of data showing some discontinuities in the solar wind is given in Figure
5.14.

Burlaga and Ness have suggested three scales of observing time in which
to classify interplanetary phenomena: (1) microstructure, $t \leq 1$ hours; (2)

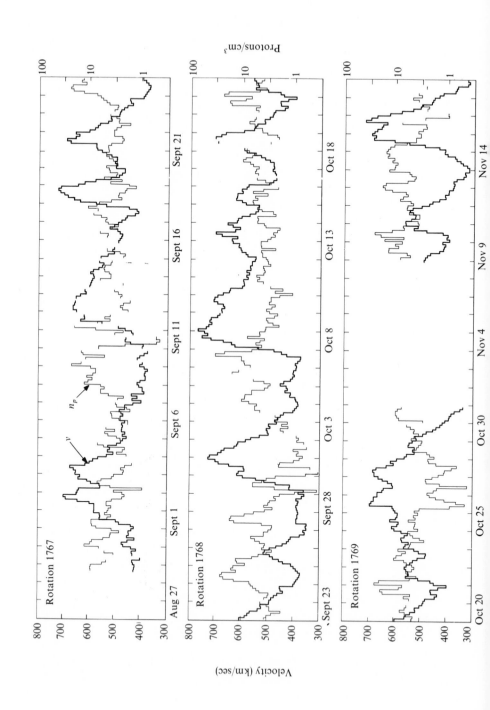

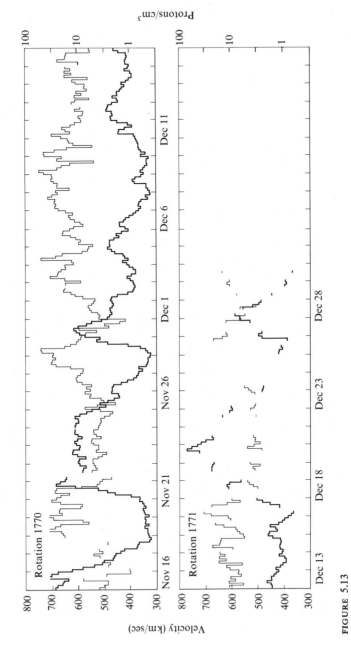

**FIGURE 5.13**

The variation of density and velocity with time as measured by Mariner 2. (Courtsey M. Neugebauer and C. W. Snyder, and the *Journal of Geophysical Research*.)

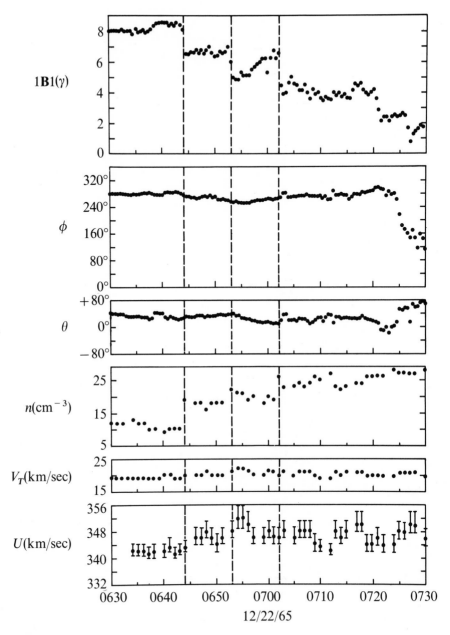

FIGURE 5.14

Sample data showing discontinuities in the solar wind; $U$ is the bulk velocity, $V_T$ is the thermal velocity, $\phi$ is the azimuthal angle shown in Figure 5.11, and $\theta$ is the polar angle ($\theta = 0$ in the ecliptic); $\phi$ and $\theta$ refer to the direction of $B$. (Courtesy L. F. Burlaga.)

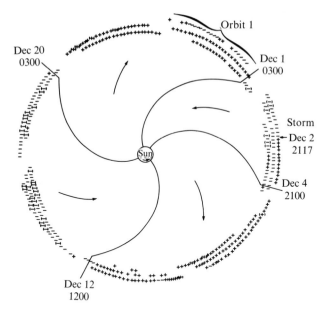

FIGURE 5.15

The sector structure of the interplanetary magnetic field
observed by IMP-1. The plus or minus polarities correspond to
the positive and negative directions indicated in Figure 5.11.
Polarities in parentheses correspond to a movement into the
shaded area of Figure 5.11 for a few hours in a smooth and
continuous manner. (Courtesy N. F. Ness and J. M. Wilcox.)

mesostructure, $1h \le t \le 10^2$ hours; and (3) macrostructure, $t > 10^2$ hours.
These observing times scale naturally into length scales by multiplying them
by an average solar wind velocity of about 400 km/sec. Then, we find: (1)
microstructure, $l(\mu s) < 10^6$ km $\approx 0.01$ a.u.; (2) mesostructure, $l(\text{meso}) \sim 10^6$–
$10^8$ km $\sim 0.01$–1.0 a.u.; and (3) macrostructure, $l(\text{macro}) > 10^8$ km $\approx 1$ a.u.

The microstructure corresponds to shock waves and the contact discontinuities; these features can act as scattering centers for cosmic rays. The
mesostructure corresponds to filaments or kinks in the field structure. These
are the "flux tubes" responsible for channeling of cosmic rays (Section 6.7).
The ultimate origin of this scale of structure may be in the solar supergranulation described in Section 2.2.

Finally, the macrostructure manifests itself as a longitudinal variation
in the magnetic field first discovered by Ness and Wilcox which they called
the "sector structure." Individual sectors contain a magnetic field of constant
polarity as shown in Figure 5.15. The sector boundaries are sharp ($< 10^5$ km)
and relatively stable in time. Changes do occur, however. Regular variations
of magnetic field magnitude, $w$, $Kp$, and $N_e$ are found in the individual

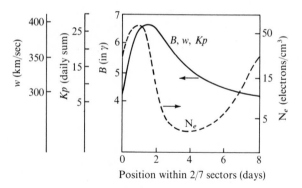

FIGURE 5.16

A schematic representation of the
variation of $N_e$, $w$, $B$, and $Kp$
through a (two-sevenths) sector.

sectors; these are shown schematically in Figure 5.16 while individual
measurements for the $Kp$ variation are shown in Figure 5.17 in order to
show the spread in the data. The sector structure also shows in the cosmic
ray observations (see Section 6.7). When the sectors were first discovered,
three sectors each occupied two-sevenths of a solar rotation while one sector
occupied only one-seventh of a solar rotation. The variations presented in
Figures 5.16 and 5.17 refer to position within a two-sevenths sector. The preceding portion of the sector (in the sense of rotation) is more active than its
following portion. The entire sector structure co-rotates with the sun (recall
the phonograph record and needle analogy presented in Section 3.1). The
discovery of the sector structure is extremely important. Organization of
solar wind properties on such a large scale was not generally expected although
Ahluwalia and Dessler had suggested substantial unidirectional field regions.

The synchronous variation of $w$, $B_I$, and $Kp$ suggests a possible statistical
relationship which could be used to calculate longterm properties of $w$ and
$B_I$ from existing records of $Kp$. The first such relationship was found by
Snyder, Neugebauer, and Rao from the Mariner 2 data, viz.,

$$w = 8.44 \, \Sigma Kp + 330 \quad (\text{km/sec}) \tag{5.5}$$

Here, $\Sigma Kp$ is the daily sum of $Kp$. Similarly, Wilcox, Schatten, and Ness
found

$$Kp \approx \frac{|B_I|}{3} \tag{5.6}$$

where $B_I$ is the interplanetary field in $\gamma$s. These relationships are statistical
in nature, and the spread in individual values is large. The exact numerical

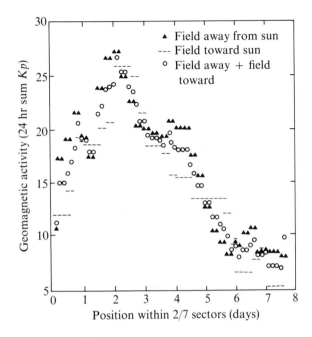

FIGURE 5.17

Actual data for the variation of $\Sigma Kp$ (twenty-four-hour sum) with position in the (two-sevenths) sector. See also Figure 1.3. (Courtesy J. M. Wilcox and N. F. Ness.)

form of these relationships should not, therefore, be taken too seriously. Other forms using other geomagnetic indices have been suggested. However, the qualitative variations are well established. There is no generally accepted physical argument for these correlations although Ballif *et al.*, have shown that $Kp$ correlates well with transverse fluctuations in the interplanetary magnetic field (Section 4.4).

The large-scale ordering of the interplanetary field suggests the possibility of a similar ordering of the photospheric field. A comparison of the extrapolated sector structure with the observed large-scale photospheric field is shown in Figure 5.18; the strong fields associated with sunspots are not recorded on this chart. One can compute the cross-correlation as a function of time between the photospheric field and the interplanetary field at earth, and there is a huge peak at $4.5 \pm 0.5$ days, as shown in Figure 5.19. Thus, approximately 4.5 days after a given magnetic region undergoes central meridian passage, a similar magnetic field passes the earth. This time delay corresponds to a constant velocity of propagation of $385 \pm 45$ km/sec, a value which is consistent with direct determinations of the plasma velocity. The velocity is certainly not constant, and the value should be regarded as

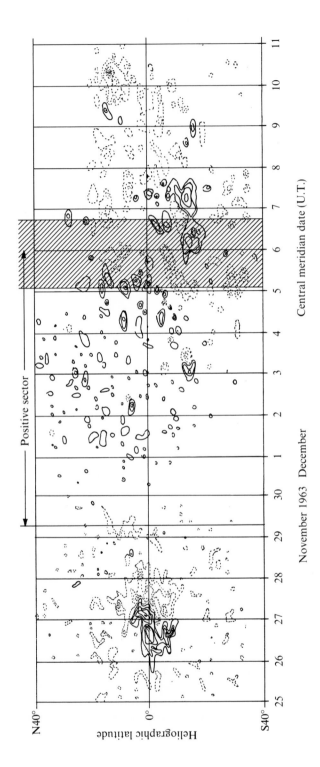

FIGURE 5.18

The synoptic chart of the photospheric magnetic field for central meridian dates November 25, 1963, to December 11, 1963 (a rectangular equal area projection), and the extrapolated position of the appropriate positive sector. Solid lines represent positive polarity, dashed lines negative polarity, and the contours are for fields of 2, 4, 8, 12, and 25 gauss. Strong fields associated with sunspots are not shown. The cross-hatched area indicates the uncertainty in the sector boundary caused by the presence of IMP-1 near perigee (and hence within the magnetosphere). (Courtesy J. M. Wilcox and N. F. Ness.)

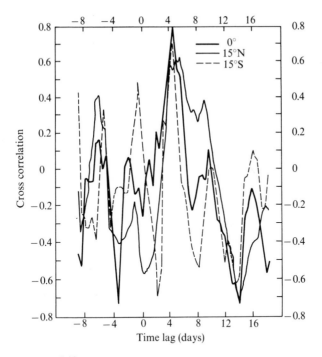

FIGURE 5.19

The cross-correlation coefficient of the interplanetary magnetic field direction and the photospheric magnetic field direction in late 1963-early 1964. Note the peak at 4.5 days. (Courtesy N. F. Ness and J. M. Wilcox.)

illustrative. In fact, Schatten, Wilcox, and Ness have calculated the field on a "source" surface at 0.6 $R_\odot$ above the photosphere. This surface is just above many of the complications near the solar surface, and the resulting fields show a better correlation with the interplanetary field than does the surface field. It takes approximately one month between the appearance of a new magnetic region in the photosphere and its appearance on the source surface at 0.6 $R_\odot$ above the surface; hence, it is suggested that new photospheric magnetic features do not influence the interplanetary field for about one solar rotation.

The variations in the photospheric field with latitude and the solar differential rotation can be used to hunt for a dominant source latitude. For example, using the "source" surface method, the source of the interplanetary magnetic field observed near the earth during the last half of 1965 appeared to be poleward of 25° solar latitude.

Since the radial field varies as $r^{-2}$ and the azimuthal field as $r^{-1}$, observations at the orbit of the earth can be extrapolated to construct a picture of the interplanetary magnetic field. Schatten, Ness, and Wilcox appear to have observed the birth and expansion of a new sector from October 1966 to January 1967. One month after the appearance of a new solar active region, magnetic loops were observed; these lines of force left the photosphere, extended into interplanetary space, and then returned to the sun. Such loops must be convected outward. A new sector was found during the next solar rotation. The disappearance of a sector has also been observed.

## 5.7    Plasma Probes and Magnetometers: Physical Principles

The plasma probes and magnetometers that are used to make space measurements may be unfamiliar to the reader. Therefore, we will review their physical principles and also the principal satellites on which the various devices were flown.

*Plasma Probes.* The words "plasma probe" are generally used to denote instruments designed to measure properties of plasma with particle energies in the range $10^2$–$10^4$ ev. A Faraday cup is shown schematically in Figure 5.20. The collector electrode lies below four grids, which are kept at the voltages shown at the right-hand side of the figure. The suppressor grid $G_1$ is kept at $-100$ volts to suppress electrons in the flow and photoelectrons from the collector plate which could be numerous if observations are made toward the sun (the normal mode of operation). A photocurrent from $G_1$ to $C$ could be significant, but it will not be modulated by the square-wave input to the modulator $G_3$. Grids $G_2$ and $G_4$ are usually kept at zero volts (spacecraft ground), and $G_2$ serves as a shield for the collector. Grid $G_1$ is called the outer grid. Sometimes $G_4$ is kept at a negative voltage. Electrons flowing into the cup are absorbed onto the grids. Protons flowing into the cup with energy less than the square wave voltage $+V$ applied to $G_3$ are modulated with the square wave frequency. This produces a square-wave current from the collector with the frequency of the modulating voltage. The current is then proportional to the proton flux entering the Faraday cup. Bulk motions can be separated from "thermal" motions if the satellite is rotating (by observing the flux in more than one direction), and thus a value of $wN_p$ is obtained. The energy distribution is obtained by varying the square wave voltage applied to $G_3$. If the incoming particles were concentrated near 1 kev, a square-wave voltage of (say) 100 volts would produce little or no modulation; a voltage of 1.5 kev would produce a dramatic modulation while a voltage of 2 kev would produce approximately the same modulation as the 1.5 kev voltage. Thus, the energy and, therefore, the velocity of the particles can be determined, and when combined with the flux determination, both $w$ and $N_p$

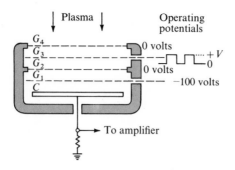

FIGURE 5.20
Schematic of a Faraday cup plasma
detector.

are determinable. The distribution of flux with energy also determines a
quantity analogous to "temperature."

This method, used in the early flights and still used when the total flux is
very low, illustrates the principles of the Faraday cup. In current usage, grid
$G_3$ is usually modulated with a voltage that changes between $V_1$ and $V_2$.
Only the current from the collector at the modulating frequency is recorded.
By successively changing the values of $V_1$ and $V_2$, the energy spectrum of the
incident flux can be determined.

Faraday cups generally suffer from poor angular resolution and photoelectric effects from the walls and grids, and several minutes are required to
obtain a complete spectrum. These devices were used, for example, on Explorer 10, IMP-1, Mariner 4, and Pioneer 6.

Curved plate or electrostatic analyzers are also commonly used. These
consist of two concentric spherical surfaces, lunes, or hemispheres ($r$ = average
radius) with a small space between and a voltage ($V$) maintained between the
two. A particle entering the space between the plates with just the right energy
per unit charge allows a balance between the electrostatic forces ($\propto QV$)
and the centrifugal forces ($mw^2/r$); such a particle with a small range of
$E/Q$ passes to the end to the exit aperture where it is counted or detected.
Incoming particles that fall outside the permitted range of $E/Q$ are absorbed
on the walls and not counted. Figure 5.21 shows the Vela 2 electrostatic
analyzer; the Vela 2 analyzer has fourteen steps in energy per unit charge
while the Vela 3 analyzer, which is similar, has sixty-four such energy steps,
and several minutes are required to obtain a complete spectrum. Histograms
of $E/Q$ obtained with such instruments are given in Figures 5.3 and 5.8.

Electrostatic analyzers can be made with good angular resolution. If used
on a rotating satellite, they can provide directional data. Besides the Vela
satellites, electrostatic analyzers have been used, for example, on Mariner 2,
IMP-1, and Pioneer 6.

The electrostatic analyzer is not a true mass spectrometer, as is desired in

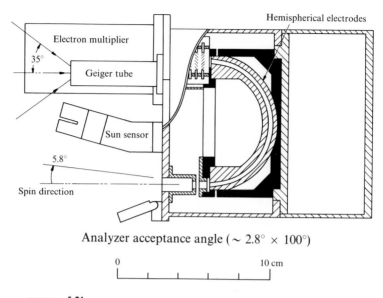

Vela 2 electrostatic analyzer

FIGURE 5.21

The Vela 2 electrostatic analyzer. (Courtesy I. B. Strong, J. R. Asbridge, S. J. Bame, and A. J. Hundhausen.)

many circumstances (see, for example, Figure 5.8). It is of considerable importance in the study of chemical abundances in the solar wind to have, in addition, a time-of-flight feature to determine directly the velocity and, therefore, the mass per unit charge ($M/Q$). Such measurements have already been made by Ogilvie, Burlaga, and Wilkerson.

*Magnetometers.* The simplest magnetometer used in space measurement is the search or spin coil, which consists of a coil of wire that is fixed to the spacecraft and spins with it. The coil axis is generally inclined to the spin axis. As the satellite spins, the coil cuts the magnetic field lines and generates a sinusoidal response with the amplitude of the emf proportional to $(dB/dt)$ sin $\theta$, where $\theta$ is the angle between the spin axis of the coil and the field lines. In essence, it is the field perpendicular to the spin axis that is measured. Such a measurement is unaffected by spacecraft fields as long as they remain constant. Two disadvantages of the search coil are: (1) it is not an absolute instrument and must be calibrated before flight and (2) the frequency of the signal is low (the spin frequency of the spacecraft). Although a search coil magnetometer was used on Pioneer 5, it is no longer a common practice to use it because the total field is not measured and because a spinning spacecraft is required.

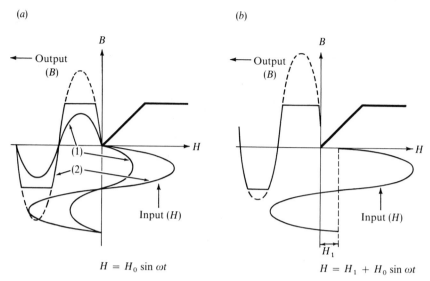

(a)

B

Output
(B)

H

(1)

(2)

Input (H)

$H = H_0 \sin \omega t$

(b)

B

Output
(B)

H

Input (H)

$H_1$

$H = H_1 + H_0 \sin \omega t$

FIGURE 5.22

Hysteresis curves illustrating the principle of the fluxgate magnetometer; the two cases shown are (a) without and (b) with ambient field.

Fluxgate magnetometers are widely used in space measurements of magnetic fields; the physical principle of saturation of a ferromagnetic core in the presence of an ambient field can be used to produce an asymmetry in the magnetic induction when the core is driven by a sinusoidal impressed field. The basic idea is shown schematically in Figure 5.22. In (a), the ambient field is zero and the impressed field is $H = H_0 \sin \omega t$. The thick solid line is the hysteresis curve which shows a linear increase of $B$ with increasing $H$ up to saturation. The input $H$ is shown along the negative $y$ axis and the output $B$ along the negative $x$ axis. For curve (1) (which does not drive the core into saturation), a simple sine wave output is obtained; such curves are characterized by "half-wave symmetry" defined by $f(t) = -f(t + \pi/\omega)$, i.e., $f(t)$ differs only by sign for parts of the curve 180° out of phase. Even for curve (2) which drives the core into saturation, half-wave symmetry is maintained. An expansion of a wave with half-wave symmetry in a Fourier series contains only odd harmonics.

This situation is altered by the presence of an ambient field as shown in (b). Here the total impressed field is $H = H_1 + H_0 \sin \omega t$, and, as can be seen, the half-wave symmetry is destroyed by the presence of the ambient field $H_1$. The Fourier expansion of such a wave form will contain even harmonics. This principle can be exploited, as shown in Figure 5.23. The driving oscillator is run at a frequency $f$ (between 5–20 kc/sec), and a signal at frequency $2f$ appears in the secondary coil only if an ambient field is present.

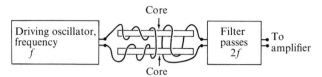

FIGURE 5.23
Schematic arrangement of a typical fluxgate magnetometer.

The use of two cores with reverse winding (as shown) cancels out the odd harmonics and the secondary feeds a filter which passes only the second harmonic. The amplitude of the signal when in use is proportional to the ambient magnetic field. The fluxgate magnetometer is not an absolute device, but it has a very large range; it is sensitive to $0.1\gamma$ and can measure fields up to thousands of $\gamma$. The sign of the field can also be determined from the phase. The fluxgate cores are small (the size of a cigarette) and can be flipped by $180°$ to provide calibration checks.

Fluxgate magnetometers have been flown, for example, on Mariner 2, IMP-1, and Pioneer 6. If the satellite is stabilized, triaxial cores must be used to determine the vector field. If the satellite is spinning, one probe (core) inclined at an angle $54° 44'$ $[=\arccos(1/3^{1/2})]$ to the spin axis can be used to determine the vector field from measurements made in different orientations. Fluxgate magnetometers are relatively insensitive to temperature changes and are often used in conjunction with rubidium-vapor magnetometers, which provide an absolute measurement.

A general class of "precession magnetometers" is based on the interaction of the nuclear magnetic moment with the magnetic field; this produces a precession of the magnetic moment about the field lines at the Larmor frequency. There are several varieties that use the same basic principle, but the most common one is the aforementioned rubidium-vapor magnetometer.

First, consider the energy level diagram for rubidium-87 as shown in Figure 5.24; the goal is to optically pump atoms into a metastable level whose energy depends on the ambient magnetic field. If a cloud of rubidium vapor is illuminated by a collimated beam with a narrow range of wavelengths centered on 7,947.6 Å and passed through a polarizer-quarter-wave plate to produce circular polarization, excitation from the $^2S_{1/2}$ state to the $^2P_{1/2}$ state occurs. In the presence of a magnetic field and the circularly polarized light, only transitions with $m = +1$ are allowed; hence, upward transitions occur from all lower levels except the one with $m = +2$. The $^2P_{1/2}$ levels have lifetimes $\sim 10^{-8}$ sec, and, hence, excitation is immediately followed by spontaneous emission to the $^2S_{1/2}$ state. Atoms back in the $^2S_{1/2}$ state but not at the $m = +2$ level are recycled. Since the $m = +2$ level is metastable, eventually all atoms in the cell (cloud) are at the $m = +2$ level. When this happens, the cell no longer absorbs the radiation at 7,947.6 Å, and the cell is transparent; this fact is easily detected by a photocell.

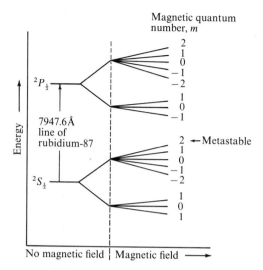

FIGURE 5.24

Schematic energy-level diagram of the two lowest
energy states of rubidium-87, illustrating the
principle of the alkali vapor magnetometer.

The metastable level $^2S_{1/2}$, $m = 2$ can be de-excited by the process of stimulated emission if an electromagnetic wave is applied perpendicular to the ambient field with just the right energy. The energy difference between magnetically split levels corresponds to the Larmor frequency of the atom, which for rubidium-87 is

$$f_L = 6.99632 \pm 0.00237 \text{ c.p.s.}/\gamma \qquad (5.7)$$

Application of electromagnetic waves with the correct energy causes a depopulation of the $m = +2$ level and causes the cell to become opaque. This event is easily detected by the photocell, the frequency is known, and the magnetic field is measured. In practice, feedback is established between the photocell and the source of the de-exciting radiation, and the transparency of the cell is made to oscillate at the Larmor frequency. For such a system, the field to be measured must be inclined to an angle to the axis of the pumping light. In addition, precession magnetometers provide no information concerning the field direction. Hence, rubidium-87 magnetometers are often flown in conjunction with fluxgate magnetometers. Alkali-vapor magnetometers have a useful range of $10\gamma$ to 1 gauss. A significant disadvantage is their relatively high power requirement.

Rubidium-vapor magnetometers have been flown, for example, on the EGO and the early IMP series of satellites. A variety of other precession magnetometers are possible, such as the helium magnetometer; these have been used on Mariner 4 and 5.

148

## 5.8 Compendium of Space Probes and Results

Many different deep space probes and satellites have contributed to our knowledge of solar wind plasma and magnetic fields; a few of the early developments have been mentioned in Section 1.3. Some of the principal satellites, deep space probes, and series are:

(1) Mariner Series: Provided extensive measurements of plasma properties ($N_e$, $w$, and $T$) and the magnetic field as well as evidence concerning the heliocentric variation.
(2) Interplanetary Monitoring Platforms (IMPs): Extensive measurements of $N_e$, $w$, $T$, and $B$; such data led to discovery of sector structure. IMP-1 used all of the major instrumental techniques described in Section 5.7; a photograph of IMP-1 is shown in Figure 5.25.
(3) Vela Satellite Series: Extensive measurements of standard quantities ($N_e$, $w$) as well as proton distribution functions leading to temperatures

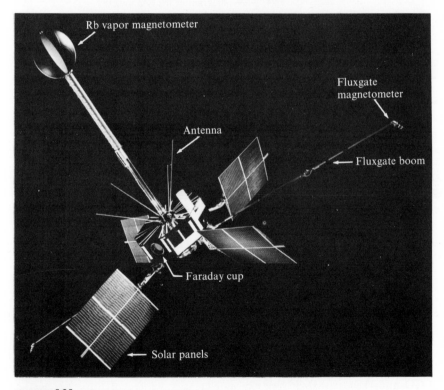

FIGURE 5.25

The IMP-1 satellite with its variety of instruments. The electrostatic analyzer is not visible. (Courtesy NASA Goddard Space Flight Center.)

and asymmetries. Also provided extensive abundance determinations and the first spacecraft measurement of $w_\phi$.

(4) Pioneer Probes: Extensive magnetic field and plasma information. Pioneer 6 provided evidence that the magnetic field component perpendicular to the ecliptic is near zero.

The launch data, orbit, and solar wind instrumentation of these satellites are summarized in Table 5.2. Orbital characteristics given in $R_E$ refer to

TABLE 5.2
*Summary of Solar Wind Probes*

| Satellite | Launch date | Orbit | Lifetime[a] (in days) | Instrumentation |
|---|---|---|---|---|
| Mariner 2 | 8–27–62 | Venus flyby 0.7–1.0 a.u. | 128 | Electrostatic analyzer, Fluxgate magnetometer |
| Mariner 4 | 11–28–64 | Mars flyby 1.0–1.5 a.u. | 270 | Faraday cup, Helium vapor magnetometer |
| Mariner 5 | 6–14–67 | Venus flyby 0.7–1.0 a.u. | 160 | |
| IMP-1 | 11–27–63 | Apogee $= 32\ R_E$ | 181 | Electrostatic analyzer, Faraday cup, Rubidium vapor magnetometer, Fluxgate magnetometer |
| IMP-2 | 10–4–64 | Apogee $= 16\ R_E$ | 150 | |
| IMP-3 | 5–29–65 | Apogee $= 42\ R_E$ | 720 | |
| IMP-4 | 5–24–67 | Apogee $= 34\ R_E$ | 710 | Electrostatic analyzer, Fluxgate magnetometer, Rubidium vapor magnetometer |
| Explorer 33 | 7–1–66 | Apogee $= 80\ R_E$ | S.O. | Faraday cup, Fluxgate magnetometer |
| Explorer 35 | 7–19–67 | In lunar orbit | S.O. | |
| Vela 2A, 2B | 7–17–64 | $r = 17\ R_E$ | S.O. | Electrostatic analyzer (on all), Search coil magnetometer (on 2A, 2B, 3A, 3B) |
| Vela 3A, 3B | 7–20–65 | $r = 18.5\ R_E$ | S.O. | |
| Vela 4A, 4B | 4–28–67 | $r = 17\ R_E$ | S.O. | |
| Pioneer 5 | 3–11–60 | Solar orbit 0.9–1.0 a.u. | 50 | Search coil magnetometer |
| Pioneer 6 | 12–16–65 | Solar orbit 0.81–0.98 a.u. | S.O. | Electrostatic analyser, Faraday cup, Fluxgate magnetometer |
| Pioneer 7 | 8–17–66 | Solar orbit 1.0–1.1 a.u. | S.O. | |
| Pioneer 8 | 12–13–67 | Solar orbit 1.0–1.05 a.u. | S.O. | Electrostatic analyzer, Fluxgate magnetometer |
| Pioneer 9 | 11–8–68 | Solar orbit 0.8–1.15 a.u. | S.O. | |

SOURCE: After Ness (1968).
[a]S.O. = Still operating (at least partially) as of September, 1969.

earth orbit; the Vela satellites are launched in pairs into nearly circular orbits. Note that the earth's magnetosphere (Section 6.3) extends to about 14 $R_E$ along the earth-sun line and to larger distances in other directions. Obviously, the probe in question must be outside the magnetosphere to make direct measurements of solar wind properties; Asbridge, Bame, and Strong have presented evidence indicating that the solar wind plasma may sense the earth's presence outside of the shock. Theoretically, this is difficult to understand because the flow speed far exceeds any speed of propagation for a disturbance. Perhaps, particles are accelerated in the boundary of the magnetosphere and injected into the solar wind.

TABLE 5.3
*Properties of Solar Wind Plasma Determined by Spacecraft*

| Quantity | Minimum | Maximum | Average |
|---|---|---|---|
| Flux $N_p w$ (ions/cm²-sec) | $10^8$ | $10^{10}$ | $2-3 \times 10^8$ |
| Velocity, $w$(km/sec) | 200 | 900 | 400–500 |
| Density, $N_e \approx N_p$ (ions/cm³) | 0.4 | 80 | 5 |
| Temperature, $T_p$(°K) | $5 \times 10^3$ | $1 \times 10^6$ | $2 \times 10^5$ |
| Thermal anisotropy ($T_{max}/T_{average}$) | 1.0 (isotropic) | 2.5 | 1.4 |
| Helium abundance, $N(He)/N(p)$ | 0 | 0.25 | 0.05 |
| Flow direction | $\pm 15°$ of radius vector; average from $\approx 2°$ east | | |
| Magnetic field, $B(\gamma)$ | 0.25 | 40 | 6 |
| B direction | Polar component variable, average in plane of ecliptic–solar equator; planar component variable, near the earth, average spiral angle $\phi \approx 45°$ | | |
| Alfvén speed (km/sec) | 30 | 150 | 60 |

SOURCE: After Ness (1968).

The scientific results are summarized in Table 5.3, which refers to properties near the earth observed during the interval 1962–1966. The "true" average properties may differ somewhat from the values chosen. Hundhausen has noted that the future thrust of solar wind observations will probably be away from bulk properties and toward fundamental problems of plasma physics, but continuous observation of the basic properties of the solar wind plasma and the solar wind field for at least two sunspot cycles will be very important for future solar wind investigations.

# Bibliographical Notes: Chapter 5

*General References*

The following cover observations of the solar wind.

1. Ness, N. F.: in Annals of the IQSY, **4**, *Solar-Terrestrial Physics: Solar Aspects*, M.I.T. Press (1969), p. 88.
2. Lüst, R. : in *Solar Terrestrial Physics*, J. W. King and W. S. Newman, Eds., Pergamon Press, New York (1967), p. 1.
3. Axford, W. I.: *Space Sci. Revs.*, **8**, 331 (1968).
4. Ness, N. F.: *Ann. Rev. Astron. Astrophys.*, **6**, 79 (1968).
5. Wilcox, J. M.: *Space Sci. Revs.*, **8**, 258 (1968).
6. Hundhausen, A. J.: *Space Sci. Revs.*, **8**, 690 (1968).

The Mariner 2 observations were first presented in:

7. Neugebauer, M., and C. W. Snyder: *Science*, **138**, 1095 (1962).

*Section 5.1*

See the general references. The radar determinations of electron density are from:

8. Koehler, R. L.: Stanford Electronics Laboratory, "Interplanetary Electron Content Measured Between Earth and the Pioneer VI and VII Spacecraft Using Radio Propagation Effects," SU-SEL-67-051 (1967).
9. Howard, H. T., and R. L. Koehler: in *The Zodiacal Light and the Interplanetary Medium*, J. L. Weinberg, Ed., NASA SP-150, Washington, D.C. (1967), p. 361.

*In situ* observations of electrons are found, for example, in:

10. Montgomery, M. D., S. J. Bame, and A. J. Hundhausen: *J. Geophys. Res.*, **73**, 4999 (1968).

See also:

11. Burlaga, L. F.: *Solar Phys.*, **4**, 67 (1968).

*Section 5.2*

Consult the general references and:

12. Hundhausen, A. J., J. R. Asbridge, S. J. Bame, and I. B. Strong; *J. Geophys. Res.*, **72**, 1979 (1967).
13. Strong, I. B., J. R. Asbridge, S. J. Bame, and A. J. Hundhausen: in *The Zodiacal Light and the Interplanetary Medium*, J. L. Weinberg, Ed., NASA SP-150, Washington, D.C. (1967), p. 365.

*Section 5.3*

See the general references and:

14. Hundhausen, A. J., S. J. Bame, and N. F. Ness: *J. Geophys. Res.*, **72**, 5265 (1967).

15. Malitson, H. H., J. K. Alexander, and R. G. Stone: *A. J.*, **73**, S69 (1968).

*Section* 5.4
Consult the general references, Reference 13, References 47 to 49 in Chapter Two, and:
16. Bame, S. J., A. J. Hundhausen, J. R. Asbridge, and I. B. Strong; *Phys. Rev. Letters*, **20**, 393 (1968).
17. Hundhausen, A. J., H. E. Gilbert, and S. J. Bame: *J. Geophys. Res.*, **73**, 5485 (1968).
18. Ogilvie, K. W., L. F. Burlaga, and T. D. Wilkerson: *J. Geophys. Res.*, **73**, 6809 (1968).

*Section* 5.5
This section is well covered in the general references.

*Section* 5.6
See the general references and:
19. Neugebauer, M., and C. W. Snyder: *J. Geophys. Res.*, **71**, 4469 (1966).
20. Burlaga, L. F., and N. F. Ness: *Can. J. Phys.*, **46**, S962 (1968).
21. Coleman, P. J., E. J. Smith, L. Davis, and D. E. Jones: *J. Geophys. Res.*, **74**, 2826 (1969).

Also see Reference 12 and Reference 36 in Chapter Six.

The various aspects of the sector structure are thoroughly covered in the general references, Reference 3 of Chapter Three, and:
22. Ness, N. F., and J. M. Wilcox: *Phys. Rev. Letters*, **13**, 461 (1964).
23. Ness, N. F., and J. M. Wilcox: *Science*, **148**, 1592 (1965).
24. Wilcox, J. M., and N. F. Ness: *J. Geophys. Res.*, **70**, 5793 (1965).
25. Ness, N. F., and J. M. Wilcox, *Ap. J.*, **143**, 23 (1966).
26. Ness, N. F., and J. M. Wilcox: *Solar Phys.*, **2**, 351 (1967).
27. Wilcox, J. M., K. H. Schatten and N. F. Ness: *J. Geophys. Res.*, **72**, 19 (1967).
28. Schatten, K. H., J. M. Wilcox and N. F. Ness: *Solar Phys.*, **6**, 442 (1969).
29. Schatten, K. H., N. F. Ness, and J. M. Wilcox: *Solar Phys.*, **5**, 240 (1968).
30. Schatten, K. H.: Thesis (Physics Department), University of California, Berkeley (1968).

*Section* 5.7
Scientific spacecraft and instrumentation are covered in:
31. Corliss, W. R.: *Scientific Satellites*, NASA SP-133, Washington, D.C. (1967).
32. Richter, H. S., Ed.: *Instruments and Spacecraft*, NASA SP-3028, Washington, D.C. (1966).

*Section* 5.8
Consult the general references, in particular Reference 1, and:
33. Asbridge, J. R., S. J. Bame, and I. B. Strong: *J. Geophys. Res.*, **73**, 5777 (1968).

# 6

# *Interactions in the Solar System*

As the solar wind flows through the solar system, it interacts with the comets, the moon, and the planets as well as with the interplanetary dust and cosmic rays. In many cases, the existence and properties of the solar wind are essential to an understanding of the solar system object.

## 6.1   Comets

The historical use and importance of comets in the study of the solar wind have been described in Sections 1.1, 1.3, and 4.1. Such studies were concerned primarily with the motions of knots or kinks in the ionic tails. Here we will discuss the effect of the solar wind on the basic structure of comets.

The major parts of a comet are the nucleus, the coma, and the tail. The nucleus is presumed to be a solid body with a radius in the range 1–10 km and an estimated and very uncertain mass in the range $\sim 10^{16} - 10^{21}$ grams. It is postulated that the nature of the nucleus corresponds to Whipple's "icy-conglomerate" model; relatively complex parent compounds such as $H_2O$, $NH_3$, $CH_4$, $CO_2$, $C_2N_2$, etc., are found in frozen form as well as meteoric material of a wide range of sizes. These parent compounds sublimate in vacuum at temperatures of a few hundred °K; the parent molecules, when dissociated, provide the daughter molecules that are observed in cometary spectra. Cometary nuclei are known to have survived fairly close gravitation encounters without breakup and, hence, must possess considerable cohesiveness. Model calculations for a cometary nucleus indicate a surface temperature in the range 150–250 °K. The nucleus and the coma are often called the "head" of a comet.

The coma is a spherical volume (centered on the nucleus) of dust and neutral molecules expanding at a rate of about 0.5 km/sec. The process leading to this expansion of the coma is analogous to the origin of the solar wind (see Section 7.5). Typical dimensions for a detectable coma are a radius of $10^5$–$10^6$ km from the nucleus. Spectroscopically, the molecules CN, CH, NH, OH, $C_3$, and $NH_2$ have been identified in the coma. This list could be incomplete because the observations are limited to the visible part of the spectrum. The coma is visible during most of the perihelion passage of a comet.

The tails are the most spectacular feature of comets and can reach up to 1 a.u. in length; ionic comet tails appear to require a solar wind to explain them properly. The ionic comet tails are referred to as Type I and are composed predominantly of $CO^+$ with some contribution from $N_2^+$, $CO_2^+$, $CH^+$, and $OH^+$. The Type I tails are generally long and straight and point approximately radially away from the sun (Section 4.1). Type I tails are filled with fine structure such as thin filaments, knots, and kinks. The tail itself seems to be composed of many tail rays which lengthen and turn in time toward the tail axis. The tail rays extend into and appear to originate in a volume confined to within $\sim 10^3$ km of the sunward size of the nucleus.

Type II tails are smooth and fairly homogeneous; they are usually curved and are probably composed of dust particles with a typical dimension $\sim 1\mu$. Type I and Type II tails can exist separately or simultaneously. The general appearance of a large comet, such as Comet Mrkos, is illustrated in Figure 4.1.

The repulsion of cometary material into the tail is a fundamental part of comet physics and was long considered the role of solar radiation pressure (Section 1.3). At present, only Type II tails seem to require the solar radiation pressure. In 1957 Alfvén suggested the basic form for the interaction of a comet with the solar wind. According to his investigations, neutral cometary molecules (essentially at rest with respect to the solar wind) are ionized by charge-exchange with solar wind protons or by solar ultra-violet radiation. The newly produced ions are captured onto the solar wind magnetic field lines; this "loading" slows down the plasma near the head and causes the field lines to wrap around and turn to the tail axis. Axford noted that a shock transition should occur similar to the earth's standing bow shock (Section 6.3). Processes in the shock could produce energetic electrons, which could increase the ionization, etc.

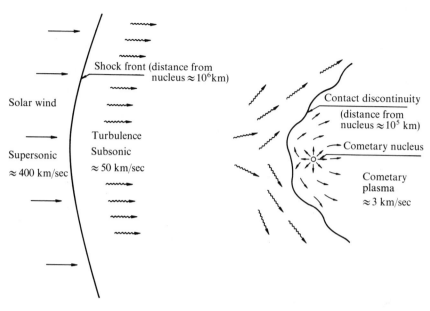

Idealized comet-solar wind interaction. (Courtesy L. Biermann, B. Brosowski, and H. U. Schmidt.)

The literature on this subject is extensive and diverse; we describe here a quantitative model investigated by Biermann, Brosowski, and Schmidt. Axial symmetry is assumed with respect to the sun-comet line, and the model applies only to the sunward side of the comet. It is also assumed that a flow of neutral molecules emanates from the nucleus and that the molecules move radially at constant velocity as long as they remain un-ionized. In these initial calculations, only photoionization is considered. When ionized, the cometary molecules are quickly accelerated to the mean flow velocity of the surrounding plasma. Thus, the equations describing the flow of the plasma component contain a source term, and mass is not conserved. Supersonic flows are not stable to additions of heavier particles, and a detached shock develops when the mean molecular weight of the solar wind flow increases by an amount corresponding to a 1 percent mixture of $CO^+$ ions by number. The ionization time for CO in the interplanetary space (either by photoionization or charge-exchange) is $\sim 10^6$ sec. Hence, we expect the detached shock to occur at a distance $\sim 10^6$ km corresponding to travel at typical ejection velocities ($\sim 1$ km/sec) before ionization becomes important. Also found in the solution is a stagnation point $10^5$ km from the nucleus; this is indicative of the contact discontinuity which separates the mixed solar wind and cometary plasma from the pure cometary plasma. This model is shown schematically in Figure 6.1.

Biermann, Brosowski, and Schmidt have themselves noted the preliminary nature of this model; for example, the observations indicate a much closer confinement of the cometary plasma to the nucleus. Perhaps inclusion of additional sources of ionization will improve the comparison between theory and observation. The strength of the source of neutral gas (molecules/sec) is also a possible cause of error. However, the source strength is also a parameter in the theory of Type II tails (the gas outflow carries the dust particle flow to their terminal velocity), and approximately the same value is indicated. Thus, the model shown in Figure 6.1 is only a first start and much work remains to be done.

Type II tails could interact with the solar wind in two ways. The first is rather indirect. For some comets, Type I and Type II tails appear to be tangent near the nucleus, particularly when viewed on plates taken with short exposure. This tangency seems to result from a drag force exerted on the dust particles by the plasma in the Type I tail. Presumably, the direction and nature of the ionic plasma flow is ultimately determined by the properties of the solar wind.

The second way concerns fine structure, which is usually not a pronounced feature of Type II or dust tails, but does exist. The so-called "synchronic band" structure is occasionally observed as striations near the end of the tail. No satisfactory explanation of this phenomenon exists within traditional theories that are based solely on radiation pressure. Dust particles may become charged through the solar ultraviolet and the photoelectric effect. This opens the possibility of collective, hydromagnetic phenomena, such as waves. Helical or corkscrew-like features in the dust tail of Comet Ikeya-Seki are suggestive of magnetohydrodynamic forces.

In addition, the "activity" of comets as evidenced by the number and brightness of tail knots ($CO^+$ clouds), etc., appears to correlate with solar activity and geomagnetic activity when the geometry is favorable. This fact was cited years ago by Biermann in connection with Comet Halley. No satisfactory explanation is available in terms of a comet model, but probably enhanced streams of solar wind flow are responsible. For example, outbursts (in brightness) of periodic comet Schwassmann-Wachmann 1 ($5.5$ a.u. $\leq r \leq 7.3$ a.u.) seem clearly correlated with M-region geomagnetic disturbances and the solar streams (sector structure) responsible for them.

## 6.2    Interplanetary Dust and the Zodiacal Light

The zodiacal light is caused by the existence of micron-sized dust particles in the interplanetary medium; the dust is in orbit around the sun and shows some concentration toward the plane of the ecliptic. The smallest particles that are not blown out of the solar system by the solar radiation pressure have radii of about $0.3\mu$ ($= 3 \times 10^{-5}$ cm). The dust originates from comets and/or from collisions between asteroids in the asteroid belt. The dust grains probably are charged by the photoelectric effect of the sun's ultraviolet radiation. When the appropriate calculation is carried out (as has been done by several authors), charges of 5–10 volts are found.

The solar wind interacts with the interplanetary dust in a variety of physical processes. The solar wind contributes to the Poynting-Robertson effect, which has been traditionally discussed in terms of solar radiation photons scattered off a dust particle carrying off the momentum of the particle; a similar effect can occur for solar wind protons. As the effect operates, particles spiral in toward the sun. However, they do not appear to reach the sun because, as they near it, vaporization of the particle occurs and the ratio of radiation pressure to solar gravity increases. The particle is then ejected from the solar system.

The so-called "corpuscular pressure" of the solar wind for a spherical particle with radius $a$ is

$$P_c = (\pi a^2)k(N_p w^2)m_p \tag{6.1}$$

where $k$ is an efficiency factor. The radiation pressure $P_r$ is given by the analog of equation (1.10) for dust particles. With these expressions, the tangential Poynting-Robertson force is

$$F_{PR} = -\left(\frac{P_r}{c} + \frac{P_c}{w}\right)u \tag{6.2}$$

where $u$ is the tangential velocity of the dust particle. A characteristic time is readily estimated from equation (6.2) and is

$$t_{PR} = \frac{M}{\dfrac{P_r}{c} + \dfrac{P_c}{w}} \tag{6.3}$$

where $M$ is the mass of the particle. Schmidt and Elsässer have computed some lifetimes for dielectric particles ($\rho = 1$) and $k = 1$; these results are given in Table 6.1. The solar wind flux was taken to be $10^9$ electrons/cm²-sec; this value is a little high, but could be representative if flare events are included.

The influence of the solar wind is dramatic and dominates the removal of small particles for $a \leq 10^{-5}$ cm. Its effect is comparable with the radiation pressure for the larger particles ($a \sim 10^{-4}$ cm) thought to be responsible for the zodiacal light.

The solar wind also interacts with dust particles through the mechanism of "sputtering," or the continual and progressive abrasion of the surface layers

TABLE 6.1
*Poynting-Robertson Lifetimes*

| a(cm) | $t_{PR}$ (years) | |
| --- | --- | --- |
| | Radiation only | Total |
| $10^{-7}$ | $3.4 \times 10^8$ | 2.4 |
| $10^{-5}$ | 340 | 140 |
| $10^{-3}$ | $7 \times 10^3$ | $6 \times 10^3$ |

SOURCE: After Schmidt and Elsässer (1967).

of the particles by ion collisions. This mechanism is difficult to evaluate quantitatively, but it appears that a disruption $\approx 10 m_p$ per ion is reasonable. This immediately leads to a mass loss of $\sim 10^{-14}$ g/cm²-sec. If this is applied to a particle with $a = 10^{-4}$cm and $\rho = 2.5$ g/cm³, such a mass loss leads to a decrease in radius $\sim 1$ Å/yr. Thus, the time scale for destruction by sputtering is $\sim 10^3 - 10^4$ years for typical zodiacal light particles.

Finally, the solar wind interacts with the dust particles through their charge and the electric field in the plasma. Recall [equation (3.51) and the discussion immediately preceding it] that the electric field seen by an observer moving with the plasma is zero, or

$$E_1(\text{plasma}) = E + v \times B = 0 \tag{6.4}$$

Since the orbital speed of the dust is much less than the solar wind speed, $v$, the dust particles see an electric field

$$E_1 (\text{dust}) = E = -v \times B \tag{6.5}$$

Such a field acting on a charged dust particle has interesting consequences.

If there is a persistent $B_\theta$ in one direction as was reported from early space magnetometer results (Section 5.5), an $E$ in the plane of the solar equator would result. Estimates of lifetimes for dust particles under this force are very small. For example, a spherical 1-micron particle in a circular orbit at 1 a.u. would have a lifetime of only thirty years. Such a lifetime would pose grave problems in the maintenance of the zodiacal light and would certainly imply large brightness fluctuations. Since the solar implications of a persistent $B_\theta$ are immense (Section 5.5) and later field measurements do not show it, there is no point in pursuing this subject any further.

However, the fields in the plane of the solar equator have an alternating effect because of the sector structure (Section 5.6). The electric force $(w \times B)$ points upward (say) with respect to the solar equator in one sector and then downward, etc. During the passage of one sector, a velocity increase $\sim 0.1$ km/sec could be expected. If the various increases add statistically, a 1-micron particle would reach escape velocity in about $3 \times 10^3$ years. This time scale is comparable with the others discussed here.

Thus, the solar wind has several important and distinct effects on the interplanetary dust particles that produce the zodiacal light. Particles with $a \leq 10^{-5}$ cm are rapidly removed. For the typical particle with $a \approx 10^{-4}$ cm, the solar wind effects reduce the lifetime by a factor $\approx 3$ to 4 compared with the lifetime for the Poynting-Robertson effect with radiation alone; hence, the lifetime is $\approx 10^3$ years.

## 6.3   The Earth

The solar wind interacts with the earth's magnetic field and produces a large variety of phenomena including parts of geomagnetic activity and the aurorae. Entire monographs have been written on this subject, and only the broad outlines are presented here.

The earth's magnetic field can be represented with reasonable accuracy by a centered dipole, viz.,

$$B = \frac{M_E}{r^3}(1 + 3\sin^2 \lambda_m)^{1/2} \tag{6.6}$$

where $r$ is the geocentric distance, $M_E$ is the magnetic moment ($8 \times 10^{25}$ gauss-cm$^3$ corresponding to $B \approx \frac{1}{2}$ gauss on the earth's surface), and $\lambda_m$ is the magnetic latitude. The magnetic axis is inclined to the rotation axis, and the magnetic north pole is at latitude 78° N, longitude 69° W (near Thule, Greenland). The equation for a line of force can be written as

$$r = r_1 \cos^2 \lambda_m \tag{6.7}$$

where $r_1$ is the distance to the field line at the magnetic equator. Field lines shaped according to equation (6.7) produce the familiar dipole appearance, which applies near the surface of the earth. However, the magnetic pressure in the field falls off as $r^{-6}$ and eventually becomes comparable with the directed pressure in the solar wind. The terrestrial field cannot maintain its dipole character and is therefore, terminated by the solar wind.

The general location of the termination of the geomagnetic field, or the magnetopause, can be estimated from a pressure balance between the solar wind and the geomagnetic field. It is often assumed that the solar wind particles are elastically reflected from the magnetopause. This yields

$$\frac{B_c^2}{8\pi} = 2N_e \, m_H \, w^2 \cos^2 \theta \tag{6.8}$$

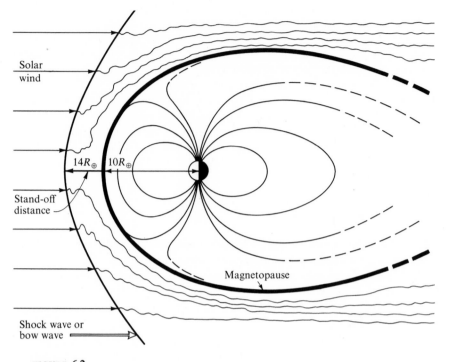

Solar wind

$14R_\oplus$   $10R_\oplus$

Stand-off distance

Magnetopause

Shock wave or bow wave

FIGURE 6.2

Schematic of the interaction between the solar wind and the terrestrial magnetic field. (From *Solar System Astrophysics*, by J. C. Brandt and P. W. Hodge, McGraw-Hill Book Co., New York, 1964.)

where the solar wind notation is standard, $B_c$ is the terrestrial field at the boundary (cutoff), and $\theta$ is the angle made by the point on the boundary and the sun-earth line as seen from the center of the earth. This problem is complex because the solar wind modifies the field in the outer parts of the magnetosphere by compression. Thus, the boundary and the precise conditions which fix it are both to be determined. Available calculations indicate a compression of a factor of about 2.6 in $B$ on the sunward side. We can solve for the cutoff distance and find

$$r_c = \left[ \frac{(2.6)^2 M_E^2}{16\pi N_e m_H w^2} \right]^{1/6} \tag{6.9}$$

For $N_e = 4/\text{cm}^3$, and $w = 5 \times 10^7$ cm/sec, we obtain $r_c \approx 10\ R_\oplus$. Because of the sixth-root variation in equation (6.9), large changes in parameters would be necessary to change this value significantly.

A standing shock front or bow wave would be expected because the terrestrial field is an obstacle in a supersonic or super-Alfvénic flow; a transiton to subsonic flow is necessary for the solar wind to flow smoothly around the

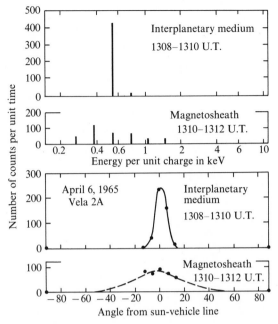

FIGURE 6.3

Changes in the plasma properties across the bow shock.
The plasma bulk velocity decreases, the "temperature"
increases, and the velocities become more isotropic.
(Courtesy J. T. Gosling, J. R. Asbridge, S. J. Bame,
and I. B. Strong.)

earth. This can be seen from the requirement of a zero velocity of the solar
wind normal to the magnetosphere. Clearly, a supersonic solar wind cannot
receive knowledge of the obstacle ahead and fulfill this requirement. The wind
can fulfill this requirement via a shock transition to subsonic flow. The general
situation is shown schematically in Figure 6.2. Outside of the shock, we
expect essentially undisturbed solar wind (but see Section 5.8). Between the
shock and the magnetopause is a region called the magnetosheath, which
consists of shocked solar wind plasma and a disordered magnetic field; the
change is clearly shown in Figure 6.3. The solar wind plasma and magnetic
field generally do not penetrate into the magnetosphere. The magnetopause
boundary appears to be stable (see Section 3.10).

The standoff distance in aerodynamic flow is given by

$$\Delta = 1.1D \frac{(\gamma - 1)}{(\gamma + 1)} \tag{6.10}$$

where $\gamma$ is the ratio of specific heats and $D$ is the effective radius of the front
of the magnetosphere. A value of $D \approx 13\ R_{\oplus}$ is representative, and with

## IMP-1 shock wave and magnetopause traversals

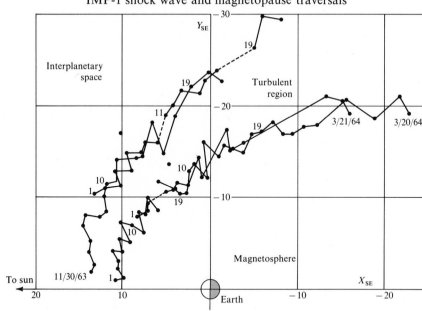

FIGURE 6.4

Determinations of the position of the bow shock and the magnetopause from the IMP-1 magnetometer. There is good agreement with theory. (Courtesy N. F. Ness.)

$\gamma = 5/3$, we find $\Delta = 3.6\ R_\oplus$. This places the bow shock at a distance of about 13–14 $R_\oplus$ in essential agreement with the observations.

The overall picture described here has been checked experimentally many times by numerous spacecraft crossings of the shock and the magnetopause. The observations are given in Figure 6.4. The agreement with theory is excellent, but some variation in the positions of the shock and the magneto-pause is indicated. Observations indicate a magnetopause thickness of $\approx 100$ km. This value is comparable with the cyclotron radius for a proton just inside the magnetopause. For 1 kev protons and $B = 70\gamma$, the cyclotron radius is 65 km.

The flow of material after the shock is such that it is accelerated again from subsonic to supersonic flow. The effect of the flow in the magnetosheath and the general interaction with the solar wind sweeps the polar field lines into a "magnetic tail." The physics and extent of this tail are not simple. Observations made on board Explorer 33 indicate a tail significantly longer than 80 $R_\oplus$ and a length of several hundred earth-radii is indicated. Such a magnetic tail can be compared with the ionic tails observed on comets.

## 6.4   The Moon

The solar wind interaction with the moon appears to be entirely different from the interaction with the earth because the moon appears to have little or no intrinsic field. The solar wind particles are directly incident on the lunar surface and are absorbed there. It appears that the absorption of the solar wind protons and alpha particles is responsible for the very low albedo of the moon (Bond albedo = 0.073). Many powdered rock materials darken under exposure to proton beams with energies ~1 kev as shown by Hapke. The lunar albedo is reached after an exposure equivalent to $10^5$–$10^6$ years. Additional exposure actually lowers the albedo below the lunar value. Thus, the observed value is probably a balance of darkening by solar wind exposure and lightening by ultraviolet bleaching and the overturn of new unexposed material by meteoritic bombardment.

   The absorption of the solar wind on the lunar surface creates a cavity in the solar wind. In addition, no shock is formed. Ness has reported that the solar wind magnetic field in the plasma cavity is relatively undisturbed and that the field lines (including discontinuities!) presumably pass through the interior of the moon. A useful quantity in interpreting these observations is a critical value of the conductivity obtained by equating the time taken for field lines to diffuse out of the moon (or be dragged through the moon) with the time required for the solar wind to flow past the moon. Recall our discussion of the magnetic Reynolds number defined by equation (3.18). The critical conductivity is obtained by setting the magnetic Reynolds number equal to unity and identifying the characteristic length as the radius of the moon, viz.,

$$\sigma_c = \frac{1}{4\pi w R_{\text{☽}}} \tag{6.11}$$

For the moon and $w = 4 \times 10^7$ cm/sec, we have $\sigma_c \approx 10^{-17}$ (c.g.s.), or $\approx 10^{-6}$ mho/meter in practical units. This value is an upper limit for the conductivity of the lunar interior. If it were higher, the field lines could not be dragged through the moon quickly enough and they would pile up in front of the moon causing a disturbance. This observationally determined upper limit to the conductivity of the lunar interior allows limits to be placed on the composition and the temperature.

## 6.5   Venus and Mars

Measurements of the intrinsic magnetic fields of Venus and Mars by flyby spacecraft indicate that their magnetic moment is very small. Hence, the solar wind interaction is with the solid planet or the atmosphere. The radii of Venus and Mars, and the moon, are not greatly different, and, hence, the critical value of $\sigma_c \approx 10^{-6}$ mho/meter quoted in Section 6.4 is applicable if the interaction is with the solid planet.

However, the atmospheres of Venus and Mars greatly alter the interaction. If the solar wind encounters an atmospheric layer, the critical conductivity is given by equation (6.11) with the planetary radius replaced by a typical atmospheric thickness such as a scale height H. For Mars a typical scale height is $\approx 30$ km, so that H $\approx 0.01\ R_\delta$. Thus, $\sigma_c \approx 10^{-4}$ mho/meter for a pileup of interplanetary magnetic field as the result of a Martian atmosphere. Such conductivities are not found in neutral (un-ionized) atmospheric layers, but they are found in ionospheric layers. Typical conductivities in the terrestrial $E$ and $F$ region are $2 \times 10^{-2}$ mho/meter and 20 mho/meter, respectively. Thus, similar ionospheres on Mars (and Venus) would effectively stop the solar wind; on the other hand, the solar wind would exert a pressure on the ionosphere and tend to push it into the lower atmosphere. Since the ionosphere is an obstacle to the solar wind, a bow shock forms at a standoff distance estimated at 0.4 $R_\delta$. The general picture sketched by Dessler is shown in Figure 6.5. In this picture, the shocked solar wind plasma in the magnetosheath can directly impinge on parts of the Martian atmosphere.

Dessler has emphasized a different physical aspect of the situation. If the solar wind (with magnetic field) blows through a conducting medium placed in the wind, a current is induced. The sense of the current is to augment the solar wind field on the windward side and decrease it on the leeward side. One can solve for a "critical current" [corresponding to the critical conductivity in equation (6.11)] that is necessary for disruption of the solar wind flow and then ask if such currents lead to acceptable current densities in the Martian atmosphere. An affirmative answer is generally possible only for an ionosphere.

At present, no observations on Mars are available to test the theoretical picture originally derived for Mars. However, measurements made during the October 1967 Mariner 5 Venus flyby indicate that the physical picture (described for Mars) applies to Venus. Measurements of plasma and the magnetic field as well as of the radio occultation are consistent with the picture shown in Figure 6.5. The standoff distance could be close to 0.4 $R_\varphi$, as expected for scaling of the terrestrial results. The downward region is a cavity of decreased density. The Mariner 5 investigators have suggested a general name of "anemopause" for the surface which separates the solar wind plasma from the planetary atmosphere, wake, or magnetic field. Anemopause, from the greek, is literally "where the wind stops." Magnetopause, while appropriate for the terrestrial case, is inappropriate for Venus and probably inappropriate for Mars.

Thus, the solar wind interaction with the moon, earth, and Venus all are fundamentally different in three ways: (1) the solar wind is absorbed on the

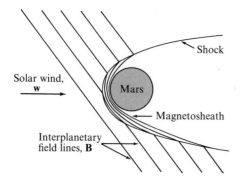

FIGURE 6.5
Schematic of the solar wind interaction with Mars.
(After A. J. Dessler.)

lunar surface and no shock develops. The conductivity of the lunar interior is low enough to permit the interplanetary field to pass through the moon without significant change. A small cavity forms behind the moon; (2) the solar wind encounters the terrestrial magnetic field as an obstacle. A magnetopause, magnetosheath, bow shock, and magnetic tail develop; and (3) the solar wind encounters the ionosphere of Venus as an obstacle and forms an anemopause and bow shock. The cavity behind Venus is filled in further downward.

## 6.6 Interaction with Jupiter

Radio observations of Jupiter indicate a population of energetic particles in a Jovian magnetosphere. The solar wind is undoubtedly the ultimate energy source.

The Jovian decimetric radiation (typical frequency of 1000 Mc/s) comes from an area roughly 6 $R_{2\!\!\!|}$ in length along the equator and 2 $R_{2\!\!\!|}$ in width along the poles. The source is linearly polarized, and the maximum polarization occurs in the outer equatorial regions. The rocking of the plane of polarization indicates a source in a magnetic field (probably dipole) with symmetry axis inclined $\approx 8°$ to the axis of rotation.

The decametric radiation (typical frequencies of 20 to 30 Mc/sec) occurs in "bursts" lasting from a fraction of a second to tens of seconds. Circular and elliptical polarizations are observed. Emissions at frequencies greater than 30 Mc/sec occur only at very restricted longitude ranges. Moreover, the emission only occurs when Jupiter's satellite Io is in the correct longitudinal range.

All theories of the Jovian decimetric and decametric radiation appear to depend on the existence of a Jovian magnetosphere and belts of trapped radiation, analogous to the terrestrial van Allen belts. For the decametric bursts, Io appears to exert a disturbing influence; this has interesting consequences concerning the internal composition of Io. Thus, the ultimate energy source for the radio emission from Jupiter is the solar wind, just as it is the ultimate energy source for geomagnetic activity.

If the Jovian radio emission is triggered by the arrival of beams of solar particles, then we might expect a correlation between radio emission and solar or geomagnetic activity. No positive correlation for the decimetric and decametric radiation with the sunspot number has been accepted currently. In fact, a negative correlation is found with the sunspot number for decametric radiation. If the Jovian ionosphere is made dense or more extensive at solar maximum, the critical plasma frequency could be raised and the correlation, therefore, may be due to an increased difficulty of escape for the radio radiation.

However, the decametric bursts have, at times, seemed to be correlated with solar flares and the associated geomagnetic activity. The travel times for the particles thought to be responsible for the disturbances are two days from the sun to the earth and eight days from the earth to Jupiter. The delays correspond to constant speeds of about 900 km/sec, and such speeds are consistent with our knowledge of flare-associated events. Nevertheless, the evidence is not conclusive and should be regarded with considerable reserve. For example, there is undoubtedly considerable deceleration of most flare-induced shock waves in interplanetary space. In fact, Sastry has argued that a physical significant correlation has not yet been proven.

## 6.7   Cosmic Rays

Solar and galactic cosmic rays interact with the solar wind via the magnetic field, which channels the solar cosmic rays and modulates the galactic cosmic rays. Knowledge of these physical processes is essential to our understanding of both the field configuration near the sun and the energy density of galactic cosmic rays near the sun. Our knowledge of cosmic rays in the solar system has been greatly extended through measurements carried out on Pioneers 6 and 7 by McCracken and co-workers.

Solar flares are well-known producers of cosmic rays, and the cosmic rays reach the earth or the spacecraft with varying delay times. Cosmic rays from flares near the middle of the sun's western hemisphere have very short delay times. This is the solar position directly connected to the earth by the Archimedes spiral configuration of the magnetic field in the solar wind. Thus, if the geometry is favorable, cosmic rays from solar events are channeled by the magnetic field directly to the observer. The longer delay times found for flares with unfavorable geometry (such as those in the eastern hemisphere)

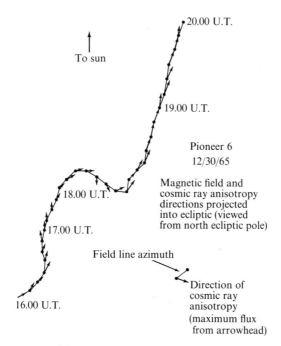

20.00 U.T.

To sun

19.00 U.T.

Pioneer 6
12/30/65

Magnetic field and
cosmic ray anisotropy
directions projected
into ecliptic (viewed
from north ecliptic pole)

18.00 U.T.

17.00 U.T.

Field line azimuth

Direction of
cosmic ray
anisotropy
(maximum flux
from arrowhead)

16.00 U.T.

FIGURE 6.6

Data showing the alignment of magnetic field and
cosmic ray anisotropy for a four-hour period on
December 30, 1965. (Courtesy K. G. McCracken
and N. F. Ness.)

can be explained by the time required for the particles to diffuse across the field lines until they reach an area with a direct connection to the observer. Most, if not all, solar flares produce cosmic radiation with energies $\leq$ 10 mev.

The detailed nature of the "channeling" or collimation described above can be dramatically illustrated by a comparison of the direction of the cosmic ray anisotropy with the direction of the magnetic field observed by McCracken and Ness on Pioneer 6 as shown in Figure 6.6; Figure 6.7 presents a schematic diagram of the filamentary structure in the interplanetary field. One can see very clearly the guiding of the particles around "kinks" in the magnetic field. The plasma properties (density, velocity, temperature) showed no anomalies as the "kink" passed the spacecraft. One should think of discrete tubes of magnetic flux extending from the sun to 1 a.u. and beyond. These tubes could have their ultimate origin in the solar supergranulation cells (Section 2.2). From studies carried out simultaneously on IMP-3 and Pioneer 6, Ness has shown a uniformity of the interplanetary magnetic field on a characteristic length of 0.01 a.u. or $10^6$ km. The observations do not indicate

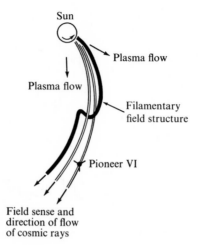

Sun

Plasma flow

Plasma flow

Filamentary
field structure

Pioneer VI

Field sense and
direction of flow
of cosmic rays

FIGURE 6.7

Schematic of the field structure
responsible for the orientations given
in Figure 6.6. (Courtesy K. G.
McCracken and N. F. Ness.)

a regular or combed out arrangement of flux tubes; rather, Burlaga and Ness described them as like "noodles on a plate."

The degree of anisotropy of the cosmic rays can be treated as follows. If the counting rates $X$ or flux of cosmic rays can be represented by an isotropic component plus a unidirectional component, then

$$X = X_0 + A \cos (\Phi - \Phi_0) \tag{6.12}$$

The anisotropy is the quantity $A$ which points in the direction $\Phi_0$; the degree of anisotropy is then given by $A/(A + X_0)$. The "equilibrium anisotropy" is approximately radial and not field-aligned. The "nonequilibrium aniso-tropy," which occurs during the early part of a flare event, is field-aligned. A sample of data (containing a Forbush decrease and a particle event) is shown in Figure 6.8. The plot shows the "equilibrium anisotropy" of 5–15 percent in addition to the higher values found during the nonequilibrium situation when the intensity of solar-generated particles is either increasing or fluctuat-ing. The disturbances responsible for Forbush decreases were discussed in Section 3.8.

The energetic particles exhibit still another characteristic, as shown in Figure 6.9. They appear somewhat confined to the magnetic flux tube on which they originate and co-rotate with the field. Figure 6.9 shows a recurrent stream found in one sector for at least three solar rotations. This implies either generation or storage that is confined to one sector. Such a phenomenon

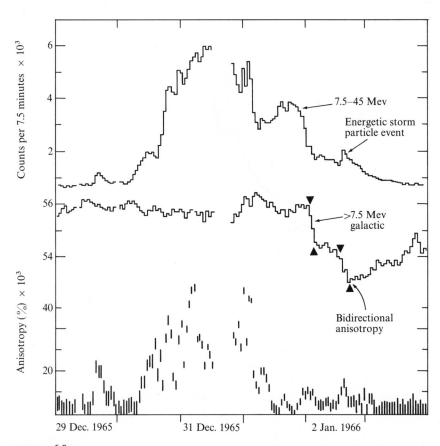

FIGURE 6.8

Cosmic ray events associated with a solar flare, viz., the increase in solar cosmic rays and the resulting (Forbush) decrease in the galactic cosmic rays. Note the increase in the anisotropy during the solar event. (Courtesy K. G. McCracken, U. R. Rao, and R. P. Bukata, and the *Journal of Geophysical Research*.)

is clearly a challenging problem of solar physics. Also note in Figure 6.9 that the usual variation through a sector is present; compare with Figure 5.16 in Section 5.6.

After flare-associated particles pass the observer, there is an increase in the intensity of particles from the anti-solar direction and eventually an approximately isotropic distribution, which then decays. Such behavior is attributed to the "trapping" of solar cosmic rays in the inner solar system by irregularities in the magnetic field lines beyond the earth. Such irregularities could originate in the plasma instabilities described in Section 3.9 and would serve as "scattering centers" for the cosmic rays. A distribution of irregularities from 2 to 5 a.u. could account for the storage and eventual leakage of the solar cosmic rays.

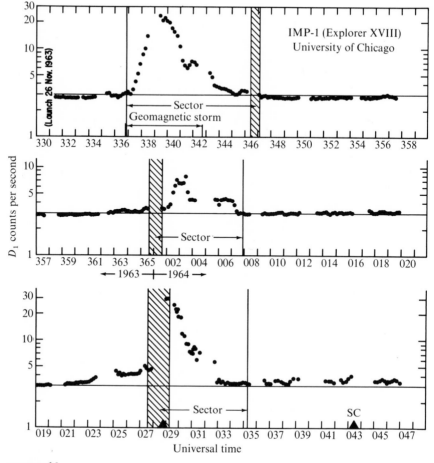

FIGURE 6.9

Recurrent streams of 1 mev protons (observed by J. A. Simpson) and the sector structure. (Courtesy J. M. Wilcox and N. F. Ness.)

These scattering centers also tend to sweep galactic cosmic rays out of the solar system, producing "solar modulation." Thus, the intensity of galactic particles in the solar system varies inversely with (say) the sunspot number, as shown in Figure 6.10. We can imagine that the extent and the magnitude of the disordered field region are larger at solar maximum, etc. Thus, the galactic cosmic rays have a harder time diffusing into the solar system at solar maximum than at solar minimum. Parker has developed a theory of cosmic ray modulation by the solar wind. However, conflicting evidence concerning, for example, the mean free path of cosmic ray particles near the earth appears to indicate the need for a more empirical approach. Measurements made by particle detectors carried to the orbit of Mars and beyond can be utilized.

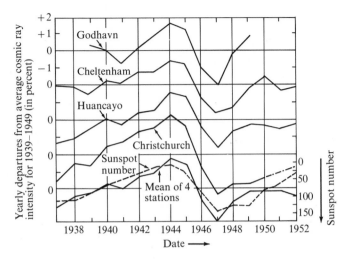

The modulation of galactic cosmic rays as indicated by the inverse correlation between sunspot number and ionization chamber intensity. (After S. E. Forbush.)

According to Parker's theory, the inward diffusion of particles (caused by a density gradient) is balanced by the outward convection of particles by magnetic irregularities in the solar wind. Thus, for equilibrium, we have

$$\frac{\partial u(r)}{\partial r} = \frac{w}{k} u(r) \tag{6.13}$$

where $w$ is the solar wind speed, $u(r)$ is the cosmic ray particle density, and $k$ is the diffusion coefficient. The value of $k$ depends on charged-particle parameters such as magnetic rigidity $R$ and velocity $v = \beta c$ as well as the properties of the interplanetary medium. The magnitude of $k$ cannot be inferred from studies at one distance, but it can, in principle, be calculated in terms of the "power spectra" of the interplanetary magnetic field (Section 5.6). Jokipii and Coleman have computed diffusion coefficients and radial gradients from the magnetic field power spectra that they determined from observations made on Mariner 4, and these compare favorably with the direct determinations on board the same spacecraft made by O'Gallagher and Simpson. Results obtained by Sari and Ness from observed magnetic power spectra are also consistent with recent observations of the modulation.

Measurements carried out by O'Gallagher and Simpson with instrumentation carried on Mariner 4 provide radial gradient information over a range of 0.56 a.u. during a time of minimum modulation. A radial gradient of $9.6 \pm 0.9$ (percent/a.u.) was found for protons with mean magnetic rigidity in the range from 4 to 7 BV. This translates into $k \approx 1.3 \times 10^{22}$ cm²/sec at $\approx 6$ BV magnetic rigidity for a solar wind speed of 400 km/sec. (Considerable caution should be used in evaluating these results because they are questionable and because some experimenters have actually observed a *negative* gradient of cosmic ray intensity with increasing distance.)

The relation between the modulated density at $r$ and the unmodulated density at $r = \infty$ can be written as

$$u(R, \beta, r, t) = u(R, \beta, \infty) \exp - \int_r^L \frac{wdr}{k(R, \beta, r, t)} \qquad (6.14)$$

where $L$ is the extent (unknown) of the modulating region. Equation (6.14) illustrates the difficulties in extrapolating beyond the modulating region to determine the density of cosmic rays in interstellar space. Not only is $L$ unknown, but so are the functional dependences for $k$, although the Mariner 4 data have circumscribed the possibilities. If the Mariner 4 gradient extends to 5 or 10 a.u., the energy density of cosmic rays in the local interstellar space would exceed the generally accepted value of $\approx 1$ ev/cm³ by a factor of two or more. Since the cosmic rays are an important contributor to the total energy density of interstellar space (kinetic energy of random cloud motions, magnetic field energy, and starlight are all approximately 1 ev/cm³) and the particles must be contained by the field, such a change in the cosmic ray energy density could have serious consequences concerning the structure and stability of the galaxy.

In addition, the Mariner 4 results indicate considerable modulation even at solar minimum where the least possible effect is expected. Some authors have assumed no modulation at solar minimum, and clearly this assumption must be discarded.

# Bibliographical Notes: Chapter 6

*Section* 6.1

A recent review of cometary phenomena is in:
1. Brandt, J. C.: *Ann. Rev. Astron. Astrophys.*, **6**, 276 (1968).

The model described in the text and extensive references are found in:
2. Biermann, L., Brosowski, B., and H. U. Schmidt: *Solar Phys.*, **1**, 254 (1967).

See also:
3. Richter, N. B.: *The Nature of Comets*, Dover Publications, New York (1963).

*Section* 6.2

The problems of the zodiacal light are covered in:
4. Weinberg, J. L., Ed.: *The Zodiacal Light and the Interplanetary Medium*, NASA SP-150, Washington, D.C. (1967).

The physics of the dust particles is reviewed and discussed in:
5. Biermann, L. : *The Zodiacal Light and the Interplanetary Medium*, J. L. Weinberg, Ed., NASA SP-150, Washington, D.C. (1967), p. 279.
6. Schmidt, T., and H. Elsässer: in *The Zodiacal Light and the Interplanetary Medium*, J. L. Weinberg, Ed., NASA SP-150, Washington, D.C. (1967), p. 287.
7. Belton, M. J. S.: in *The Zodiacal Light and the Interplanetary Medium*, J. L. Weinberg, Ed., NASA SP-150, Washington, D.C. (1967), p. 301.

*Section* 6.3

Discussions of the magnetosphere and additional references are found in:
8. Hess, W. N., and G. D. Mead: in *Introduction to Space Science*, 2nd ed., W. N. Hess and G. D. Mead, Eds., Gordon and Breach, New York (1968), p. 373.
9. Ness, N. F., C. S. Scearce, J. B. Seek, and J. M. Wilcox: *Space Res.*, **VI**, 581 (1966).
10. Spreiter, J. R., A. L. Summers, and A. Y. Alksne: *Planetary Space Sci.*, **14**, 223 (1966).
11. Brandt, J. C., and P. W. Hodge: *Solar System Astrophysics*, McGraw-Hill Book Co., New York (1964), p. 397.
12. Ness, N. F.: *Rev. Geophys.*, **7**, 97 (1969).

Additional papers on magnetospheric phenomena are found in the February 1969 issue of *Reviews of Geophysics*.

Also see:
13. Carovillano, R. L., J. F. McClay, and H. R. Radoski, Eds., *Physics of the Magnetosphere*, D. Reidel Publishing Co., Dordrecht-Holland (1968).

*Section* 6.4

The lunar albedo is discussed in:

14. Hapke, B. W.: in *The Nature of the Lunar Surface*, W. N. Hess, D. H. Menzel, and J. A. O'Keefe, Eds., The Johns Hopkins Press, Baltimore, Md. (1966), p. 141.
15. Cohen, A. J., and B. W. Hapke: *Science*, **161**, 1237 (1968).
16. Ness, N. F.: *Space Res.*, **IX**, 678 (1969).
17. Ness, N. F., K. W. Behannon, C. S. Scearce, and S. C. Cantarano: *J. Geophys. Res.*, **72**, 5769 (1968).
18. Taylor, H. E., K. W. Behannon, and N. F. Ness: *J. Geophys. Res.*, **73**, 6723 (1968).

*Section* 6.5

The theoretical discussion of the solar wind interaction with Mars is in:

19. Dessler, A. J.: in *The Atmospheres of Venus and Mars*, J. C. Brandt and M. B. McElroy, Eds., Gordon and Breach, New York (1968), p. 241.

The Mariner 5 observations of Venus are presented in consecutive papers beginning with:

20. Snyder, C. W.: *Science*, **158**, 1665 (1967).

*Section* 6.6

The Jovian radio emission is reviewed and discussed in:

21. Warwick, J. W.: *Ann. Rev. Astron. Astrophys.*, **2**, 1 (1964).
22. Dulk, G. A.: *Science*, **148**, 1585 (1965).

Correlations with solar activity are presented in:

23. Carr, T. D., A. G. Smith, H. Bollhagen, N. F. Six, and N. E. Chatterton: *Ap. J.*, **134**, 105 (1961).

A dissenting view is presented in:

24. Sastry, Ch. V.: *Planetary Space Sci.*, **16**, 1147 (1968).

*Section* 6.7

A comprehensive reference to cosmic ray physics is:

25. Sandstrom, A. E.: *Cosmic Ray Physics*, Interscience Publishers, New York (1965).

Early work is reviewed in Reference 1 of Chapter Three and in:

26. Savedoff, M. P., Ed.: *Ap. J. Suppl.*, **IV**, 369 (1960).

Cosmic ray observations from spacecraft are presented in consecutive papers beginning with:

27. McCracken, K. G., U. R. Rao, and R. P. Bukata: *J. Geophys. Res.*, **72**, 4293 (1967).

See also:

28. Burlaga, L. F.: *J. Geophys. Res.*, **72**, 4449 (1967).

The detailed results concerning collimation are presented in:

29. McCracken, K. G., and N. F. Ness: *J. Geophys. Res.*, **71**, 3315 (1966).

30. McCracken, K. G., U. R. Rao, and N. F. Ness: *J. Geophys. Res.*, **73**, 4159 (1968).

The flux tubes and related topics are discussed in the general references to Chapter Five and in:
31. Ness, N. F.: *J. Geophys. Res.*, **71**, 3319 (1966).
32. Burlaga, L. F., and N. F. Ness: *Can. J. Phys.*, **46**, S962 (1968).
33. Wilcox, J. M., and N. F. Ness: *J. Geophys. Res.*, **70**, 5793 (1965).

The modulation is discussed in Reference 1 of Chapter Three and in:
34. Forbush, S. E.: *J. Geophys. Res.*, **59**, 525 (1954).
35. Jokippi, J. R., and P. J. Coleman: *J. Geophys. Res.*, **73**, 5495 (1968).
36. Sari, J. W., and N. F. Ness, *Solar Physics*, **8**, 155 (1969).
37. O'Gallagher, J. J., and J. A. Simpson: *Ap. J.*, **147**, 819 (1967).

# 7

# Impact on Astrophysics

## 7.1 Solar Origin of the Solar Wind

The solar origin of the solar wind is a complex problem of solar physics. The probability or ease of solar wind flow "clearly" varies over the solar disk. Enhanced flow could originate in active coronal regions with higher temperatures and densities than the undisturbed corona. Reduced flow could result from regions with a high magnetic field or a closed field configuration. These situations can exist simultaneously near a solar-active region. With the aid of some simplifying assumptions, Pneuman has investigated the form of the expansion from a bipolar magnetic region. The results (Figure 7.1) resemble the helmets described in Section 2.4. Expansion is choked off in the central part where the field lines return to the sun; near the periphery, the plasma flow carries the field away from the sun.

Such considerations raise the question of an essentially general expansion or solar wind emission from discrete "solar nozzles." The nozzle hypothesis has its advantages, for example, in explaining the sector structure (Section 5.6). The fact that the solar wind magnetic field is essentially unidirectional for

178

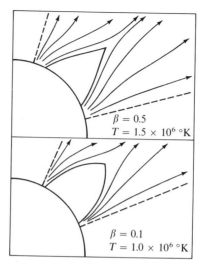

$\beta = 0.5$
$T = 1.5 \times 10^6 \ ^\circ K$

$\beta = 0.1$
$T = 1.0 \times 10^6 \ ^\circ K$

FIGURE 7.1

The form of the flow from a solar
bipolar magnetic region. (Courtesy
G. W. Pneuman.)

substantial parts of a solar rotation (e.g., about one-half a rotation in 1962) can be easily explained if the flow originated in a small solar region with a unidirectional field.

The resolution of this problem may seem to be quite simple because "only" an extrapolation back to the solar surface is required. Since the velocity is observed and theory is available, the extrapolation appears straightforward. This is a problem of long standing in solar-terrestrial relations, and, until rather recently, all such extrapolation attempts have been somewhat equivocal, for the velocity of the solar wind is very low near the sun (recall the equation of continuity), streams of different velocity can interact, nonradial motions can be important, and complicating structures can exist in the corona itself. Some of these effects can be overcome by utilizing a "source surface" (Section 5.6) located about $0.6 \ R_\odot$ above the photosphere. Extrapolations of the sector structure carried out by Ness and Wilcox appear to have been successful; these extrapolations have the tremendous advantage of a verifiable consistency check, namely, the polarity of the photospheric field. These extrapolations, however, are not sufficient in themselves to distinguish between the competing hypotheses of general expansion and discrete nozzles. In particular, evidence for motion in solar latitude continues to crop up.

Present evidence (as summarized by Wilcox) tends to favor the mapping hypothesis, in which magnetic field lines connect each solar longitude to an equivalent longitude at the orbit of the earth. The observations of energetic proton streams (Section 6.7) and the longitudinal distribution of flare events which produce fast travel times to the spacecraft or to the earth support this hypothesis; it appears unlikely that a nozzle would always be in just the right location. Moreover, the time delay of about 4.8 days (found from a positive peak in the cross-correlation curve between the photospheric field and the interplanetary field at 1 a.u.) is the result of a procedure that gives equal weight to each solar longitude. It is difficult to see how such a correlation could arise if the field from one small solar area was spread out over a sector (see Figure 5.19 in Section 5.6). Finally, the probability distribution of plage area within a sector appears to be similar to the variation of plasma velocity and the magnitude of the interplanetary field within a sector (Figure 5.16). This is consistent with the mapping or general expansion hypothesis; note that plage areas are generally associated with photospheric fields of 10 to 20 gauss or greater (see Figure 5.18).

However, the evidence for significant cosmic ray storage, the appearance of coronal features, and theoretical discussions such as Pneuman's argue strongly for solar regions with a closed magnetic field and hence no expansion. Some insight into this problem can be obtained by attempting to calculate the field configuration in the corona from the observed photospheric field distribution. The basic relation is one of Maxwell's equations

$$\mathbf{V} \times \mathbf{B} = 4\pi\mathbf{J} \tag{7.1}$$

Calculations with this equation require the specification of $\mathbf{J}$. If $\mathbf{J} = 0$, we have $\mathbf{V} \times \mathbf{B} = 0$, and the field configuration can be calculated from potential theory. A force-free assumption can be made in which $\mathbf{J}$ is parallel to $\mathbf{B}$. Then, the equation becomes

$$\mathbf{V} \times \mathbf{B} = \alpha(r)\mathbf{B} \tag{7.2}$$

where $\alpha(r)$ is constant along a line of force to ensure $\mathbf{V} \cdot \mathbf{J} = 0$. Such calculations have been carried out by Newkirk, Altschuler, and Harvey and by Schatten, Wilcox, and Ness. A sample result for the current free approximation is shown in (a) of Figure 7.2; in (b) of Figure 7.2, the influence of the solar wind is included by specifying radial field lines at a certain distance from the sun. Note that only the field configuration is significant (the density of field lines is not) and that fine structure in the magnetic field is not represented.

The appearance of the corona for the same time is shown in Figure 7.3. Many similarities exist. The coronal streamers in the southwest and southeast quadrants appear to be located over "arcades" of magnetic loops. It seems that calculation of (coronal) magnetic fields from potential theory yields valuable information below 2 $R_\odot$. The drawback is a tendency to overestimate the number of closed loops.

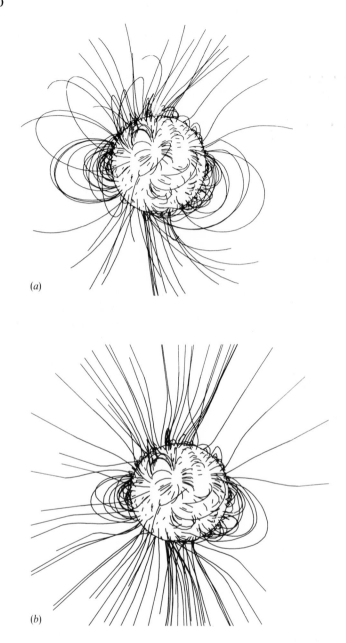

(a)

(b)

FIGURE 7.2

The projected magnetic field lines calculated for November 12, 1966, for the current free approximation (a), and for an approximation (b) that includes the influence of the solar wind by requiring radial field lines at a sufficient distance from the sun. Longitude of disk center = 270°. (Courtesy G. Newkirk, Jr., and M. D. Altschuler.)

FIGURE 7.3

The solar corona on November 12, 1966, taken with a radially symmetric, neutral density filter to compensate for the sharp decrease in coronal brightness with distance. Venus appears in the northeast quadrant. (Courtesy G. Newkirk, Jr.)

Hence, the present status definitely favors a solar wind that arises from a general expansion of the solar corona, but the "mapping hypothesis" cannot be supported in its extreme form. Certain areas of flow or emission are undoubtedly preferred or forbidden, and the fraction of the solar surface involved in the general expansion could be considered (say) as about one-half; this fraction increases as one moves away from the surface.

## 7.2 The Solar Model and General Relativity

The solar interior may appear at first sight to be a well-understood bastion of astrophysical theory. This assumption has recently been questioned, however. Predictions of the solar neutrino flux are in apparent contradiction with the observations, indicating a flaw in the observing technique or a basic deficiency in our knowledge of solar interior conditions, particularly temperature or nuclear reaction rates. Even the traditional assumption of an essentially solid-body rotation of the sun has been questioned and a rapidly rotating core suggested as an alternative. This latter question is intimately connected with the solar wind.

The sun's figure or dynamic oblateness bears directly on the astronomical tests of general relativity. The deflection of light near the solar limb observed at solar eclipse is sufficiently accurate definitely to favor general relativity over Newtonian mechanics, but for other relevant comparisons the accuracy is not high. This fact places added emphasis on the need for comparison of observation with theory for the advance of the perihelion of Mercury. The observed value of 40 sec of arc per century is in apparent agreement with observation; note, however, that substantial corrections (e.g., influences of planets) are involved. Alternate theories, such as the Brans-Dicke cosmology, predict different perihelion advances. The Brans-Dicke cosmology actually predicts a smaller perihelion advance (by $\approx 4$ sec of arc/century) than observed, but agreement could be obtained if a small advance were due to the motion of a planet revolving about a slightly oblate sun. Such an oblateness could be due to a rapidly rotating interior with a time period of about one day.

Evidence supporting this contention comes from Dicke and Goldenberg's measurements of the apparent oblateness of the solar disk in white light, viz., $\varepsilon = 5 \times 10^{-5}$ given in equation (2.1). Conversion of the apparent oblateness to the dynamical oblateness is a problem, but Dicke and Goldenberg argue that this value is essentially the dynamical oblateness; such a value would influence the advance of Mercury's perihelion to the extent that is required; it also opens up the question of the accuracy of the general relativistic tensor theory of gravitation in comparison (say) with the scalar-tensor theory of gravitation in the Brans-Dicke cosmology.

If the solar core is rotating with a period of one or two days and if the surface layers or atmosphere is rotating with the observed twenty-seven-day period, the possibility of interaction exists. Since the interior contains essentially all of the mass and angular momentum, the atmosphere will tend to accelerate to the angular velocity of the interior unless there is a braking force acting on the atmosphere. Dicke has suggested that the solar wind provides this braking. As the plasma moves away from the sun, the co-rotating field exerts an accelerating torque on the escaping plasma; this constitutes a braking torque on the solar atmosphere. The observed torque/steradian of about $7 \times 10^{29}$ dyne-cm/ster (see Sections 3.7, 4.1, and 5.2) is close to the value estimated by Dicke for a steady-state exchange of angular momentum with the interior.

Strictly speaking, this agreement amounts to a consistency check. Even if the solar interior is not in rapid rotation, the solar wind provides a natural mechanism for loss of solar angular momentum (Section 7.3). The hypothesis of a rapidly rotating solar interior is currently the subject of considerable controversy; for example, the stability of such a regime is under question. At the very least, it has stimulated considerable interest in the physics of the solar interior.

## 7.3 Solar and Stellar Rotation

The disposition of the angular momentum of the sun and stars is an astrophysical problem of long standing. Some sort of braking mechanism seems required for most stars because contraction with angular momentum conserved from protostar dimensions with very small transverse velocities leads to very large rotational velocities for stars with main sequence dimensions. Formation of binaries or magnetic coupling with the surrounding interstellar medium have been proposed to explain the relatively low velocities of stellar rotation that have been observed. The angular momentum sink associated with mass loss could also be significant.

*Solar Rotation and Braking.* As is the case in many other areas of stellar astrophysics, the sun provides us with a unique chance to determine the time rate of change of angular momentum for an individual star. Observations of solar wind parameters near the earth determine the angular momentum per unit mass $L$ in the solar wind near the solar equator; the quantity $L$ is directly related to the Alfvén point in the solar wind [see equations (3.60) and (3.63)]. If we assume that the radial velocity and density of the solar wind are independent of solar latitude $b$ and that the angular momentum per unit mass varies with latitude as $\cos^2 b$, we have

$$
\begin{aligned}
\frac{dJ_\odot}{dt} &= -4\pi r^2 m_H N_e w L(b=0) \int_0^{\pi/2} \cos^3 b \, db \\
&= -\frac{8}{3} \pi r^2 m_H N_e w L(b=0)
\end{aligned}
\tag{7.3}
$$

where $J_\odot$ is the solar angular momentum. Note also that

$$
\frac{dM_\odot}{dt} = -4\pi r^2 m_H N_e w
\tag{7.4}
$$

where $M_\odot$ is the mass of the sun. If we assume that the sun rotates as a solid body, the deceleration goes as

$$
\frac{dJ_\odot}{dt} = -\frac{J_\odot}{\tau}
\tag{7.5}
$$

where $\tau$ is the *e*-folding time for solar rotational braking. The angular momentum $J_\odot$ can be written in terms of the moment of inertia $I_\odot$ and the expression for $\tau$ becomes

$$\frac{1}{\tau} = \frac{dM_\odot}{dt} \frac{2r_A^2}{3I_\odot}$$

(7.6)

For reference, the solar angular momentum is $\approx 1.7 \times 10^{48}$ gm-cm²/sec, and the moment of inertia is $\approx 6.0 \times 10^{53}$ gm-cm². For these values and the solar wind parameters discussed in Chapters Three and Five, we find $\tau$ to be in the range $3$–$6 \times 10^9$ years. Such numbers are comparable with the currently accepted age of the sun, $5 \times 10^9$ years. Of course, if the sun resembles the model proposed by Dicke, these numbers do not apply; in this case, only a constantly increasing outer shell is braked by the solar wind. Remember that a constant solar wind is assumed in these calculations.

The sun could have had a substantially larger rotational velocity in the past of (say) 75 km/sec (see below). The original equatorial rotational velocity could have been $\approx 10$ km/sec, as implied by the time scales given above. If this is the case, additional factors must be important, such as a possible enhanced solar wind earlier in the sun's evolution.

The loss of solar angular momentum in the solar wind is a rather efficient process. During a solar lifetime of $4.5 \times 10^9$ years, the fraction of angular momentum lost is comparable with the total, but the mass loss at the present rate produces a total loss of only $10^{-4}M_\odot$. Of course, only the solar material with the largest angular momentum per unit mass is lost in the solar wind.

*Rotations of Solar-Type Stars.* Considerable insight into the rotational history of solar-type stars could be obtained by determining the rotational velocities of solar type stars of different ages. The ages could be assigned by the usual method applied to clusters (i.e., the main sequence turnoff point) and the rotational velocities derived from the broadening of line profiles. Such an investigation has been carried out by Kraft.

Kraft considered rotational velocities of F and G stars in the field, in the Hyades, and in the Pleiades, all of which have been determined from line profiles; stellar rotation cannot be distinguished from turbulent motions by an analysis of line profiles, but complications from this source are not likely to be important. Kraft determined the observational quantity $V_R \sin i$ where $V_R$ is the equatorial rotational velocity and $i$ is the inclination of the plane of the star's equator to the plane of the sky. The mean rotational velocity is given by $\langle V_R \rangle = (4/\pi)\langle V_R \sin i \rangle$.

Kraft's results indicate a strong decrease of rotational velocity with (nuclear) age. Values for main sequence stars with effective temperatures slightly larger than the sun (G0 V) and with mass $M/M_\odot = 1.2$ are as follows:

for Pleiades stars, the nuclear age is $3 \times 10^7$ years and $\langle V \rangle = 39$ km/sec; for Hyades stars, the nuclear age is $4 \times 10^8$ years and $\langle V \rangle = 6$ km/sec. Thus, the time scale for reduction of rotational velocities of young (age of Pleiades) stars is approximately equal to the age of the Hyades or $4 \times 10^8$ years. The available evidence indicates a somewhat faster deceleration for the sun. An "initial" sun with $V \approx 75$ km/sec and a deceleration time scale $\sim 10^8$ years is indicated. These numbers and the $e$-folding time for the present sun imply a more efficient braking at earlier solar epochs unless the angular momentum is "hidden" in the sun, as on Dicke's model.

Additional support for this overall view comes from Walker's study of the very young cluster NCG 2264 (age $\approx 3 \times 10^6$ years). Walker noted that, in contrast to main sequence stars, the stars later than F8 appeared to have larger rotational velocities than the earlier-type stars. He noted that this behavior would be expected if these stars were recently formed and had not lost the major part of their angular momentum.

T Tauri stars are thought to be rather young stars in the process of contracting toward the main sequence; they are often in locations with appreciable gas and dust—conditions thought to be favorable for star formation. On the basis of a spectroscopic study, Kuhi found velocities of mass ejection between 225 km/sec and 325 km/sec and a mean total outflow of some $4 \times 10^{-8}$ $M_\odot/yr$. Such large outflows could help provide the rapid deceleration time which is apparently required for a very young star.

*Main Sequence Rotations.* The run of rotations along the main sequence is known, and typical values are shown in Figure 7.4. The main feature is high values of the equatorial rotational velocity for early-type stars and low values for late-type stars; the changeover occurs near F0 corresponding to a star of mass $M/M_\odot \approx 1.4$. Schatzman pointed out that the transition in main sequence rotations occurs at approximately the same spectral type as the transition between stars with substantial subphotospheric convection zones and those with envelopes in radiative equilibrium; the extent of the hydrogen convection zone as a function of spectral type is shown schematically in Figure 7.5. Schatzman proposed that material ejected from active regions and coupled to the stars by their magnetic field could produce rotational braking. Active regions are presumably associated with chromospheres and coronas, and these parts of a star's atmosphere depend on the existence of a substantial, subsurface hydrogen convection zone (Sections 2.2 and 2.3). Thus, Schatzman's proposal naturally explained the correspondence between Figures 7.4 and 7.5, although Modisette and Higgins have proposed an alternative. They suggest that all stars have stellar winds but that the relatively high rotations found for early-type stars is due to their young ages and short main sequence lifetimes.

The physical picture has evolved into one with magnetic braking occurring

186

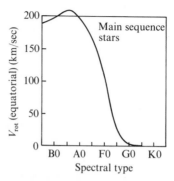

FIGURE 7.4

The variation of mean rotational
velocity along the main sequence
showing the sharp decrease at
F0 and later.

as the result of an essentially general expansion (Section 3.7). The time scale
for braking depends on the velocity and flux in the solar wind and the magnetic field [see equation (7.6)] as well as the length of time the stellar wind flows.
The time presumably corresponds to the time for existence of the hydrogen
convection zone. If the HCZ is a permanent feature, the stars continuously
lose mass and angular momentum.

As stars in their initial stages of evolution approach the main sequence, they
pass through a convective stage (discovered by Hayashi) and presumably have
mass loss. This apparently occurs for stars earlier and later than F5, and
probably all stars lose some angular momentum during their Hayashi stage.
However, as noted above, the earlier stars do not have substantial convective
zones after reaching the main sequence, and, therefore, they retain a larger
fraction of their initial rotation. The loss is continuous for the later-type stars
including the sun, and we find rotations decreasing with age (see Figure 7.4).
Thus, there seems to be substantial support for the theory that mechanical
energy is generated in a hydrogen convection zone which produces a corona
and stellar wind which, in turn, leads to substantial magnetic braking of
rotation.

The contention of active chromospheres (and presumably coronas) for
stars later than F5 can be checked by a spectroscopic search for Ca II K line
emission along the main sequence; such emission shows close correspondence
with active regions on the sun. Wilson has suggested a definition of "active
chromosphere": that which shows the emission on spectrograms of dispersion 10 Å/mm or less. According to this definition, active chromospheres
occur in stars of spectral type F5 or later. Most stars later than F5 (including
the sun) are not active by this definition.

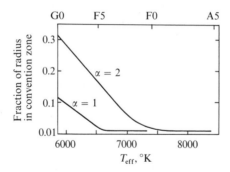

FIGURE 7.5

The depth of the hydrogen convection zone after calculations made by N. Baker and R. Kippenhahn. Results are presented for the mixing length set equal to 1 ($\alpha = 1$) and 2 ($\alpha = 2$) scale heights. (From *The Sun and Stars*, by J. C. Brandt, McGraw-Hill Book Co., New York, 1966.)

An alternative hypothesis (advocated by Huang) to explain the observed main sequence rotations ascribes the loss of angular momentum to the creation of planetary systems. Indeed, the distribution of the angular momentum in a rotating gas cloud into a star and a planetary system would result in a much reduced stellar rotational velocity. The sun and the solar system provide an interesting case study, in which the sun contains 2 percent of the angular momentum and the planets (largely Jupiter and Saturn) 98 percent. If the present sun contained all the angular momentum in the solar system, it would rotate at $\approx 10^2$ km/sec—a value comparable with the rotations of early main sequence stars.

The planetary hypothesis lacks a detailed mechanism (other than a by-product of star formation), and it is hard to imagine it as a continuous mechanism, as implied by Kraft's observations. However, there appears to be no detailed argument against the planetary hypothesis as an integral part of the formation of stars with the magnetic braking going on continuously beginning with the Hayashi stage.

## 7.4 Stellar Winds

The existence of solar-wind-type mass flows on other main sequence stars later than F5 is implicit in our discussion of rotations in Section 7.3. The existence of a substantial hydrogen convection zone would appear to imply a stellar corona and a stellar wind qualitatively similar to the solar wind. It is, however, rather difficult to go far beyond these qualitative statements. The discussion here applies only to steady, continuous mass loss. Many examples of cataclysmic mass loss are known in astrophysics.

Parker obtained some interesting results from some simple considerations. The thermal velocity of the coronal particles must be much less than the escape velocity to have a bound corona and a solar wind flow [see equation (3.5) *et sequence*]. Numerically, he obtained

$$GM_* m_H \mu / R_* kT \lesssim 4 \qquad (7.7)$$

Since $M_*/R_*$ varies only by about a factor of three along the main sequence, one would not expect a large range in coronal temperatures. The temperature should not drop much below $1 \times 10^6$ °K for a flow where energy is supplied to a thin shell at the base and carried into the plasma by conduction. Radiation becomes more efficient as a transport mechanism relative to conduction at these lower temperatures.

Also we may argue by similarity that the speeds of the stellar wind are on the order of the escape velocity, viz.,

$$w(\infty) \sim (2GM_*/R_*)^{1/2} \qquad (7.8)$$

Again since $M_*/R_*$ does not vary strongly along the main sequence, the stellar wind speeds would approximate the solar wind value. The density and flux of these stellar winds could certainly depart from the solar value(s).

The main sequence stellar winds have their ultimate origin in convective turbulence. Energy provided by dissipation of atmospheric turbulence might possibly produce similar effects. Spectroscopic evidence indicates that most giant and supergiant atmospheres are "turbulent"; many have been described with very extended chromospheres. Such stars would be prime candidates for a stellar wind flow with some physical differences from the solar case. Evidence appears to support this contention for both early- and late-type stars.

Deutsch's studies of the system $\alpha$ Herculis have been known for over two decades. The assumed geometry of the system is shown in Figure 7.6. The star $\alpha_1$ Herculis is a less luminous supergiant (M5 II) and has a projected separation of 700 a.u. from $\alpha_2$ Herculis, a G-type companion. The basic data are the equivalent width of the circumstellar Ca II H and K lines and Ca I 4,227 Å in the spectrum of the G-star as well as the Doppler shift corresponding to the projection of the expansion along the line of sight. The expansion velocity of about 10 km/sec is found. The equivalent widths give the column density of Ca I and Ca II from the observer to the G-star. This could be converted into densities via a model and a calculation of the ionization in the M-star envelope. The calculations are uncertain, but densities on the order of $10^3$ atoms/cm$^3$ are found; these correspond to mass losses $\sim 10^{-8}$ $M_\odot$/yr. This can be compared with the present solar rate $\sim 10^{-14}$ $M_\odot$/yr.

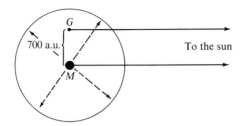

FIGURE 7.6

A schematic model of α Herculis (not to scale).
(After A. J. Deutsch.)

The temperature of the circumstellar gas associated with α Herculis is about $4 \times 10^4$ °K. Thus, conduction probably does not play the same role here as it does in the solar case. Weymann has constructed models, and the results indicate the need for a rather considerable extent of "coronal" heating. The physical situation is still uncertain, however.

Similar evidence for mass loss from O and B supergiants has been obtained by Morton and by Stecker. The spectra show absorption lines of H I, C II, C III, C IV, Si III, and Si IV displaced to the blue (shortward) by an amount corresponding to a Doppler shift from a velocity of about 2,000 km/sec; a typical spectrum is shown in Figure 7.7. Estimates of the mass ejection rate are in the range $10^{-4}$ to $10^{-6}$ $M_\odot$/yr. These rates are even larger than those found for late-type supergiants.

Thus, evidence exists for continuous mass loss with rates ranging from $10^{-14}$ $M_\odot$/yr. for the sun through $10^{-8}$ $M_\odot$/yr. for M supergiants to possibly $10^{-4}$ $M_\odot$/yr. for O and B supergiants. Such loss rates could be important in the mass balance of the interstellar medium, which is between interstellar matter going into present star formation and stellar mass being returned to the interstellar medium by steady processes (such as stellar winds) and cataclysmic processes (such as supernovae). As quantitative evidence accumulates, the stellar winds could be important in the physics of the interstellar medium. Stellar wind flows similar to the solar wind probably do not contribute significantly to the heavy element enrichment of the interstellar medium.

The influence of the mass loss in stellar winds on the structure and evolution of stars (besides the effects on rotation discussed in Section 7.3) is hard to estimate. Possible influences of mass loss on the interpretation of the Hertzsprung-Russell diagram were emphasized years ago by Soviet workers (e.g., Massevitch and Fessenkov, and Idlis). This viewpoint is no longer widely held, but the large rates of loss that are found for early-type supergiants could conceivably reopen the question. The mass loss in red giant stars could be important in terms of the Chandrasekhar limit, which requires a stellar mass less than 1.4 $M_\odot$ (a value which can be raised by rotation) in order to allow a

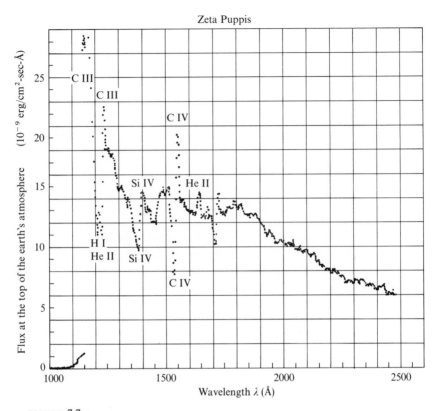

FIGURE 7.7

Spectrum of the extreme ultraviolet region of a very early 0 star (05f), Zeta Puppis. The shift described in the text can be estimated by eye from the C IV lines near 1,550 Å. The emission line comes from the star, and the absorption line comes from the expanding material between the star and the earth. The shift is about 15–20 Å; a shift of 15.5 Å corresponds to a velocity of 0.01 c or about 3,000 km/sec. (Courtesy T. P. Stecker.)

stable white dwarf. Since the white dwarf stage is widely held to be the terminal stage of stellar evolution, stars with masses greater than 1.4 $M_\odot$ must lose mass "quietly" or by a violent event. The mass loss rates for α Herculis and some OB giants and supergiants are of the right order of magnitude to produce a star with mass less than 1.4 $M_\odot$.

Finally, we note that most discussions of stellar winds assume a flow from a single star. Since some two-thirds of all stars in the galaxy are thought to be binaries, the simple theory developed in Chapter Three may not apply. Consideration of the complications has been undertaken by Nariai, and the mass loss could be enhanced in binary systems.

## 7.5 Analogous Astrophysical Applications

Solar wind or transonic type flows beginning with a subsonic velocity and passing to a supersonic flow at large distances were largely unknown in astronomy prior to 1958. Besides the de Laval or rocket nozzle and the stellar winds we have discussed, the same basic type of flow is found in other astrophysical situations.

*The Polar Wind.* The terrestrial ionosphere in the polar regions could flow along the terrestrial field lines into the magnetic tail and thence into the interplanetary space if a suitable mechanism exists. Banks and Holzer and Axford have noted similarities between the terrestrial ionosphere and the solar corona; they suggest that a polar wind of $H^+$, $He^+$, and $O^+$ may exist.

The similarities that were noted can be seen in equation (3.5) where the quantity $\lambda = (V_{esc}/U)^2$, and where $U$ is the most probable velocity. For a $2 \times 10^6$ °K corona, $\lambda \approx 4$. For the protons in the ionosphere at a height of 1,000 km with a temperature of $2 \times 10^3$ °K, $\lambda \approx 2$; a higher value holds for $He^+$. Thus, a nozzle-type flow, as outlined in Sections 3.1 and 3.3, appears possible although some controversy exists concerning the details of the flow (which has been observed).

This flow seems to be particularly important in the terrestrial helium budget in which radioactive decay of uranium and thorium in the earth's crust is balanced by escape. If the ionosphere at geomagnetic latitudes greater than 60° is considered to have an ultimate escape route, then the polar wind provides an enhanced escape flux (over the traditional exospheric treatments) and enables a balance. This treatment is preferable to the older hypothesis of high exospheric temperatures in the past (e.g., Spitzer) because it does not create problems with the escape of $^3He$.

*Comet Comas.* The flow of neutral molecules and dust particles in the coma of comets appears to be another example of transonic flow. Continuum fluid dynamics applies near the nucleus because of the short mean free path. Probstein has considered the expansion of a two-component-coupled "dusty gas," and the solution contains a critical point, marking the transition to supersonic flow. The terminal velocity for the dust grains is reached within some twenty radii of the nucleus, a distance of 20 to 100 km. For surface temperatures near 200 °K, the dust ejection velocities of 0.3 km/sec required by the widths of Type II (dust comet tails) are easily reproduced with reasonable parameters. This "inner solution" provides a starting point for detailed studies of Type II comet tails.

*A Galactic Wind.* Flows analogous to the solar wind could be important on galactic scales. Moore and Spiegel have recently proposed such a model as a description of the flow of gas from the galactic center.

## Bibliographical Notes: Chapter 7

*Section* 7.1

See References 5 and 28 in Chapter Five and:

1. Pneuman, G. W.: *Solar Phys.*, **3**, 578 (1968).
2. Newkirk, G., M. D. Altschuler, and J. Harvey: in *Structure and Development of Solar Active Regions*, K. O. Kiepenheuer, Ed., D. Reidel Publishing Co., Dordrecht-Holland (1968), p. 379.

*Section* 7.2

Consult Reference 12 in Chapter Two and:

3. Dicke, R. H.: *Nature*, **202**, 432 (1964).
4. Dicke, R. H., and P. J. Peebles: *Space Sci. Rev.*, **4**, 419 (1965).
5. Dicke, R. H.: *Ap. J.*, **149**, L121 (1967).
6. Goldreich, P., and G. Schubert: *Ap. J.*, **150**, 571 (1967).
7. Bretherton, F. P., and E. A. Spiegel: *Ap. J.*, **153**, L77 (1968).
8. Colgate, S. A.: *Ap. J.*, **153**, L81 (1968).

See also:

9. Deutsch, A. J.: *Science*, **156**, 236 (1967).

*Section* 7.3

See References 23 and 24 in Chapter Three and:

10. Brandt, J. C.: *Ap. J.*, **144**, 1221 (1966).
11. Brandt, J. C.: *Ap. J.*, **148**, 905 (1967). This paper contains a minor numerical error. $T_2$ is slightly larger such that $T_1/T_2 = 16/3\pi^2$ instead of $2/\pi$, as quoted.
12. Allen, C. W.: *Astrophysical Quantities*, 2nd ed., Athlone Press, London (1963).
13. Kraft, R. P.: *Ap. J.*, **150**, 551 (1967).
14. Walker, M. F.: *Ap. J. Suppl.*, **II**, 365 (1956).
15. Kuhi, L. V.: *Ap. J.*, **140**, 1409 (1964).
16. Brandt, J. C.: *The Sun and Stars*, McGraw-Hill Book Co., New York (1966).
17. Schatzman, E.: *The Hertzsprung-Russell Diagram*, J. L. Greenstein, Ed., I.A.U. Symposium No. 10 (1959), p. 129.
18. Schatzman, E.: *Ann. d'Ap.*, **25**, 18 (1962).
19. Wilson, O. C. : *Ap. J.*, **144**, 695 (1966).
20. Wilson, O. C.: *Science*, **151**, 1487 (1966).
21. Modisette, J. L., and P. W. Higgins: *Astrophys. Space Sci.*, in press (1969).
22. Huang, S.-S.: *Ap. J.*, **141**, 985 (1965).
23. Jastrow, R., and A. G. W. Cameron, Eds.: *Origin of the Solar System*, Academic Press, New York (1963).

*Section* 7.4

Refer to Reference 34 in Chapter Two, Reference 1 in Chapter Three, and:

24. Huang, S.-S., and O. Struve: in *Stellar Atmospheres*, J. L. Greenstein, Ed., University of Chicago Press, Chicago (1960), p. 321.
25. Biermann, L., and R. Lüst: in *Stellar Atmospheres*, J. L. Greenstein, Ed., University of Chicago Press, Chicago (1960), p. 260.
26. Deutsch, A. J.: in *Stellar Atmospheres*, J. L. Greenstein, Ed., University of Chicago Press, Chicago (1960), p. 543.
27. Deutsch, A. J.: in *Stellar Evolution*, R. F. Stein and A. G. W. Cameron, Eds., Plenum Press, New York (1966), p. 377.
28. Weymann, R.: *Ap. J.*, **136**, 476 (1962).
29. Weymann, R.: *Ap. J.*, **136**, 844 (1962).
30. Wickramasinghe, N. C., B. D. Donn, and T. P. Stecher: *Ap. J.*, **146**, 590 (1966).
31. Morton, D. C.: *Ap. J.*, **147**, 1017 (1967).
32. Morton, D. C., E. B. Jenkins, and R. C. Bohlin: *Ap. J.*, **154**, 661 (1968).
33. Stecher, T. P.: in *Wolf-Rayet Stars*, K. B. Gebbie and R. N. Thomas, Eds., National Bureau of Standards Special Publ. 307 (1968), p. 65.

The mass balance of the interstellar medium is reviewed in:

34. Burbidge, E. M., and G. R. Burbidge: *Handbuch der Physik*, **LI**, 134 (1958).

The early Soviet views on mass loss and stellar evolution are found in:

35. Massevitch, A. G.: in *The Hertzsprung-Russell Diagram*, J. L. Greenstein, Ed., I.A.U. Symposium No. 10 (1959), p. 89.
36. Fessenkov, V. G., and G. M. Idlis: in *The Hertzsprung-Russell Diagram*, J. L. Greenstein, Ed., I.A.U. Symposium No. 10 (1959), p. 115.

The problem of binaries has been considered in:

37. Nariai, K.: *Publ. Astron. Soc. Japan*, **19**, 564 (1967).

*Section* 7.5

The polar wind is discussed in:

38. Banks, P. M., and T. E. Holzer: *J. Geophys. Res.*, **73**, 6846 (1968).
39. Axford, W. I.: *J. Geophys. Res.*, **73**, 6855 (1968).

But see also:

40. Dessler, A. J., and P. A. Cloutier: *J. Geophys. Res.*, **74**, 3730 (1969).

See also the paper by P. M. Banks and T. E. Holzer following Dessler and Cloutier's article.

The theory of dust comets is discussed in:

41. Probstein, R. F.: in *Collection Dedicated to the Sixtieth Birthday of Academician L. I. Sedov*, M. A. Lavrent'ev, Ed., Izdatel'stro Akademii Nauk S.S.S.R., Moscow (1968). (English Edition, *Problems of Hydrodynamics and Continuum Mechanics*, Society for Industrial and Applied Mathematics, Philadelphia.)

See also:

42. Finson, M. L., and R. F. Probstein: *Ap. J.*, **154**, 327 (and following paper) (1968).

The galactic gas flow is discussed in:

43. Moore, D. W., and E. A. Spiegel: *Ap. J.*, **154**, 863 (1968).

# Index

Aberration, *11*
Aberration angle, *104*ff
Abundances
  coronal, *44*f, *127*
  cosmic rays, *44*f
  photospheric, *44*f, *127*
  solar wind, *127*ff, *150*
Adiabatic invariant, *125*
Alfvén critical point, *84*
Alfvén speed, *27*
Alfvénic Mach number, *83*
α Herculis, *188*f
Alpha particles, solar wind, *127*
Anemopause, *164*
Angular momentum, *83*, *84*, *183*f
Archimedes spiral, *67*, *68*, *130*
Aurorae, early work, *3*ff

Bipolar magnetic region (BMR), *48*ff
Blast wave, *90*
Bow shock. *See* Shock; Earth
Braking
  solar, *183*f
  of solar-type stars, *184*f
  stellar, *184*ff
Brightness temperature, *43*
Butterfly diagram, *48*f

Carbon cycle, *20*
Center of Activity (CA), *56*f
Chandrasekhar limit, *189*
Chapman's model, *13*f
Chemical composition. *See* Abundances
Chromosphere
  origin of, *25*ff
  stellar, *186*

Chromospheric network, *25*, *50*, *51*
Comet tails, *103*ff, *154*ff
  early ideas, *2*, *3*
  ionic, *103*ff
    knots, *11*f
    orientation, *11*, *104*ff
  Type I, *104*, *154*
  Type II, *154*, *156*, *191*
Comets
  Baade (1954h), *104*
  coma of, *154*, *191*
  Daniel (1907d), *109*
  Halley, *156*
  Haro-Chavira (1954k), *104*
  Humason (1961e), *104*
  Ikeya-Seki (1965f), *156*
  Mrkos (1957d), *105*, *154*
  nucleus of, *153*
  Periodic/Schwassmann-Wachmann 1, *156*
  Tomita-Gerber-Honda (1964c), *104*
Conductivity
  critical, *163*
    ionospheric, *164*
    lunar, *163*
  thermal, *13*, *69*
Corona
  appearance, *30*, *31*, *181*
  composition, *44*f, *127*
  density, *33*ff
  energy balance, *28*
  enhancements, *32*, *33*
  false, *33*
  fans, *30*
  flattening, *31*
  green line, *33*, *116*
  green patches, *33*